T0342523

Economic Modelling of Climate Change and Energy Policies

NEW HORIZONS IN ENVIRONMENTAL ECONOMICS

Series Editors: Wallace E. Oates, *Professor of Economics, University of Maryland, USA* and Henk Folmer, *Professor of General Economics, Wageningen University and Professor of Environmental Economics, Tilburg University, The Netherlands*

This important series is designed to make a significant contribution to the development of the principles and practices of environmental economics. It includes both theoretical and empirical work. International in scope, it addresses issues of current and future concern in both East and West and in developed and developing countries.

The main purpose of the series is to create a forum for the publication of high quality work and to show how economic analysis can make a contribution to understanding and resolving the environmental problems confronting the world in the twenty-first century.

Recent titles in the series include:

Amenities and Rural Development
Theory, Methods and Public Policy
Edited by Gary Paul Green, Steven C. Deller and David W. Marcouiller

The Evolution of Markets for Water
Theory and Practice in Australia
Edited by Jeff Bennett

Integrated Assessment and Management of Public Resources
Edited by Joseph C. Cooper, Federico Perali and Marcella Veronesi

Climate Change and the Economics of the World's Fisheries
Examples of Small Pelagic Stocks
Edited by Rögnvaldur Hannesson, Manuel Barange and Samuel F. Herrick Jr

The Theory and Practice of Environmental and Resource Economics
Essays in Honour of Karl-Gustaf Löfgren
Edited by Thomas Aronsson, Roger Axelsson and Runar Brännlund

The International Yearbook of Environmental and Resource Economics 2006/2007
A Survey of Current Issues
Edited by Tom Tietenberg and Henk Folmer

Choice Modelling and the Transfer of Environmental Values
Edited by John Rolfe and Jeff Bennett

The Impact of Climate Change on Regional Systems
A Comprehensive Analysis of California
Edited by Joel Smith and Robert Mendelsohn

Explorations in Environmental and Natural Resource Economics
Essays in Honor of Gardner M. Brown, Jr.
Edited by Robert Halvorsen and David Layton

Using Experimental Methods in Environmental and Resource Economics
Edited by John A. List

Economic Modelling of Climate Change and Energy Policies
Carlos de Miguel, Xavier Labandeira and Baltasar Manzano

Economic Modelling of Climate Change and Energy Policies

Edited by

Carlos De Miguel
Universidade de Vigo, Spain

Xavier Labandeira
Universidade de Vigo, Spain

Baltasar Manzano
Universidade de Vigo, Spain

NEW HORIZONS IN ENVIRONMENTAL ECONOMICS

Edward Elgar
Cheltenham, UK• Northampton, MA, USA

Published by
Edward Elgar Publishing Limited
Glensanda House
Montpellier Parade
Cheltenham
Glos GL50 1UA
UK

Edward Elgar Publishing, Inc.
136 West Street
Suite 202
Northampton
Massachusetts 01060
USA

A catalogue record for this book
is available from the British Library

ISBN-13: 978 1 84542 630 9
ISBN-10: 1 84542 630 4

Printed and bound in Great Britain by MPG Books Ltd, Bodmin, Cornwall

Contents

Contents

Contributors

Carlos de Miguel
rede and Department of Economic Analysis
Universidade de Vigo
Vigo, Spain

Frauke Eckermann
Chair of Public Economics
University of Dortmund
Dortmund, Germany

Tim Hoffmann
Centre for European Economic Research (ZEW)
Mannheim, Germany

Tobias Kronenberg
UNU-MERIT/Faculty of Economics and Business Administration
Maastricht University
Maastricht, The Netherlands

Xavier Labandeira
rede and Department of Applied Economics
Universidade de Vigo
Vigo, Spain

Pedro Linares
Instituto de Investigación Tecnológica
Universidad Pontificia Comillas
Madrid, Spain

Andreas Löschel
Centre for European Economic Research (ZEW)
Mannheim, Germany

Baltasar Manzano
rede and Department of Economic Analysis
Universidade de Vigo
Vigo, Spain

José M. Martín-Moreno
rede and Department of Economic Analysis
Universidade de Vigo
Vigo, Spain

Juan P. Montero
Departament of Economics
Pontificia Universidad Católica de Chile
Santiago, Chile

Ulf Moslener
Centre for European Economic Research (ZEW)
Mannheim, Germany

Sergey Paltsev
Joint Program on the Science and Policy of Global Change
Massachusetts Institute of Technology
Cambridge, USA

Agustín Pérez-Barahona
Chair Lhoist Berghmans in Environmental Economics, CORE
Université Catholique de Louvain
Louvain-la-Neuve, Belgium

Rafaela Pérez
Universidad Complutense
Madrid, Spain

Contributors

John Reilly
Joint Program on the Science and Policy of Global Change
Massachusetts Institute of Technology
Cambridge, USA

Miguel Rodríguez
rede and Department of Applied Economics
Universidade de Vigo
Vigo, Spain

Jesús Ruiz
Departament of Quantitative Economics
Universidad Complutense
Madrid, Spain

Francisco J. Santos
Instituto de Investigación Tecnológica
Universidad Pontificia Comillas
Madrid, Spain

Malte Schwoon
International Max Planck Research School on Earth System Modelling
Hamburg, Germany

Richard S.J. Tol
Department of Economics
Hamburg University
Hamburg, Germany

Sjak Smulders
Department of Economics
Tilburg University
Tilburg, The Netherlands

Mariano Ventosa
Instituto de Investigación Tecnológica
Universidad Pontificia Comillas
Madrid, Spain

Adriaan van Zon
UNU-MERIT/Faculty of Economics and Business Administration
Maastricht University
Maastricht, The Netherlands

Benteng Zou
Institute of Mathematical Economics (IMW)
Bielefeld University
Bielefeld, Germany

Acknowledgements

This book includes a selection of the chapters presented at the First Atlantic Workshop on Energy and Environmental Economics, held on the inspiring island of A Toxa in Galicia (Spain) in the late summer of 2004. The Atlantic Workshop is a bi-annual event organized by the Research group on Energy, Economics and the Environment (rede) that operates in the University of Vigo. Therefore, we are first indebted to all the members of rede for the support received before and during the organization of the workshop. The chapters and the book have also benefited from the many intellectual issues and suggestions raised by the participants in the productive and lively discussions that took place during the Atlantic Workshop.

Of course, the book would not exist without the generous financial support received by the workshop from several institutions: Caixanova, Iberdrola, the Spanish Ministry of Education (grant SEJ2004-20209E), the Galician Environmental Ministry (Xunta), Xacobeo, and the University of Vigo. Finally, we would like to thank Noelia Beceiro for a truly professional typesetting, Robert Lavigna for his careful revision of the English language, Luchi Barbosa for her always helpful administrative support and Felicity Plester for her continuous encouragement and swift responses to all our queries.

Preface

Adam Smith famously remarked: 'Man is the only animal that makes bargains; one dog does not change bones with another dog'. Europe has mobilized this unique ability of our species to trade to address the challenge of global warming. Carbon dioxide is the main greenhouse gas. For the first time since history began, beginning in January 2006, you can call your broker and buy (and sell) as many tonnes of it as you like on the European Union Emissions Trading market. The creation of this market is the most important development bar none in the battle to combat climate change. While the US waffles about technologies and partnerships, Europe puts its faith in markets. And the great irony is this. The US invented the idea of emissions trading to solve environmental problems, and insisted – over European objections – that it be included in the Kyoto Protocol. But then revisionism took hold in the US, and handouts and capture by special interests took the place of markets as the solution to the climate change challenge.

The European scheme is disarmingly simple: 6.6 billion tonnes of carbon dioxide called 'allowances' with an asset Value (@ €20 per tonne) of €132 billion have been issued to each of about 11,000 installations in the power generation and heavy industry sectors in the 25 countries of the European Union; each installation was given free a quantum of tonnes called 'allowances'. In general, utilities were left short, while the rest of heavy industry got what it needed. And the total envelope allocated by countries in the 'old' EU-15 Member States generally was less generous than allocations in the 10 new Member States. So we would expect selling from East to West, and from non utilities to utilities. Total trades documented over the counter and via exchanges amounted to about €5 billion in 2005.

The key rule in emissions trading is this; you can emit as much as you like, but at the end of the accounting period, you have to hold enough allowances to cover your emissions. If you get 500 tonnes of allowances and you emit 760 tonnes, then you have to buy an additional 260 allowances to achieve this obligation. If you get 500 tonnes, and reduce your emissions to 300 tonnes, then you can sell the surplus – 200 tonnes – in the market place and still meet your obligation. Trading produces a price per tonne of carbon dioxide which drives holders to see how they can reduce emissions and make money. This simple mechanism achieves the following; those polluters who can reduce at very low cost do so, and sell the reductions to those for whom it is relatively expensive. The overall target is achieved, and at minimum cost to the economy. The price signal releases the dynamic of innovation, as businesses everywhere try to find new and less expensive ways to reduce emissions. And it is fair – those who emit most, pay most, and *vice versa*, and the tax payer does not pick up the tab.

Business will always complain about price signals, but the best in business always responds with imagination and innovation. BP introduced its own internal emissions trading scheme in 1999; by 2001 it had reduced emissions by 10 percent, and is now an active supporter of and participant in the European scheme. And leading US companies can see the advantage of trading, as evidenced in a recent advertizement from Chevron Texaco Corporation, one of the world's largest companies: 'seeking a Carbon Markets Team manager to work in our San Ramon Ca. headquarters. This individual will coordinate carbon credits management activities to ensure that the company makes the most effective use of its carbon credit generating opportunities to satisfy its internal needs for carbon offsets'. Jeffrey Immelt, General Electric's (GE) chief executive, argues that the US government must create policies that foster renewables and address climate change: 'America is the leading consumer of energy. However, we are not the technical leader. Europe today is the major force for environmental innovation' (IEEE Spectrum on line, 3 July 2005). GE's sales of power equipment exceed $17 billion annually. Because renewables are a growing segment of the power industry, it has decided to become a leader in wind energy for electricity generation and in solar cell technology, using innovation in technology to drive the achievement of this objective. But it is doing this in spite of US policy, and it is looking abroad for markets and its developmental platform. Carbon constraint will be a continuing part of our business future. As the new generations of super carbon efficient houses, offices, cars, power stations, cement factories and associated materials and software come to market, US business will find itself at a competitive disadvantage.

The 'Vision Statement of Australia, China, India, Japan the Republic of Korea and the US for a New Asia Specific Partnership on Clean Development and Climate Change' represents the US view of the way forward. It proposes:

A new partnership to develop, deploy, and transfer cleaner, more efficient technologies and to meet national pollution reduction, energy security and climate change concerns ... A non binding compact will be developed in which the elements of this shared vision, as well as ways and means to implement it, will be further defined, and the establishment of a framework for the partnership, including institutional and financial arrangements, will be considered.

This means either nothing at all, or the provision of subsidies and associated bureaucracy to develop and transfer technologies. And we know two things: market signals are much better than subsidies and bureaucracy as a way of moving any agenda forward, especially one requiring imagination and innovation, and the Iraq War and related commitments means the US tax payer is not going to have much money to spend on new subsidies for the foreseeable future. When it comes to climate change, to paraphrase Maynard Keynes, Europe chisels in stone, while the US knits in wool.

With the agreement in Montreal in December 2005 at the United Nations Climate Change Conference to continue the global efforts to fight climate change, there is a clear implication that the European Emissions Trading Scheme will continue after 2012; it is also likely to be expanded to embrace other jurisdictions. This makes it imperative that the nuances of this scheme be studied and lessons for the future learnt, and that the interface with energy policy and energy systems be seamless and well understood. This book takes up the challenge of providing these understandings and connections. It therefore comprises an important attempt to ensure that markets continue to be at the heart of addressing climate change in Europe and globally.

December, 2005

Frank J. Convery

Heritage Trust Professor of Environmental Policy, University College Dublin

President of the European Association of Environmental and Resource Economists (EAERE)

1. Introduction and overview

Carlos de Miguel, Xavier Labandeira and Baltasar Manzano

The design and foundations of energy and (energy-related) environmental policies, together with a quantification of their effects, are very relevant from both positive and normative points of view for a number of reasons: the essential importance of energy in the functioning and survival of our societies; the growing concern caused by certain environmental problems and phenomena that are closely related to energy consumption (especially climate change); and, clearly linked to the former, the significant efficiency and distributional consequences that these policies hold for producers and consumers.

Knowing that this is a huge field, this book intends to provide an answer to only some of the previous questions and concerns. It does so by combining theoretical papers with empirical applications, policy-oriented chapters with basic contributions to the literature, and specific analysis of some policy developments with more abstract approaches. We have attempted to include some useful comprehensive papers on growth and the environment and environmental instruments, together with in-depth analysis of marginal questions related to a major interest of the book: the definition and design of public environmental and energy policies. Moreover, we stress the *ex-ante* assessment of some actual or hypothetical applications in the field. In sum, we present here a collection of chapters that contribute to the literature in several ways and could help draw a picture of some relevant aspects for public intervention in the environmental and energy domains.

To this end, the book is organized into three parts and 13 chapters. The first part deals with some fundamentals of energy and environmental policies, including this introduction and Chapters 2 and 3. Subsequently, Part II groups a number of chapters (Chapters 4 to 8) interested in different aspects of the EU Emission Trading System (ETS) already mentioned by **Frank Convery** in the

Preface to the book. Finally, the third part of the book includes Chapters 9 to 13, covering a selection of specific advanced issues pertaining to climate change and energy policies. The rest of this introduction is devoted to highlighting the main objectives and results of each chapter.

In Chapter 2, **Sjak Smulders** examines the fundamental forces that shape the interaction between growth and the environment, trying to answer very relevant questions from both scientific and policy points of view. Among them are the issues of whether it could be expected that climate change policy and associated energy policies become more (or less) stringent when the world economy grows; whether technological change would reduce the cost of climate change policy, and whether a constant carbon tax could mitigate global warming in a growing economy.

In Chapter 3, **Juan P. Montero** discusses some aspects of the performance of existing permits programs (particularly the Acid Rain Program in the US and the total suspended particulate program of Santiago), and proposals for implementation of new ones (in particular, carbon trading for dealing with global warming and a comprehensive permit program for curbing air pollution in Santiago). Furthermore, he extends the basic model of a permit market to accommodate several practical considerations such as the regulator's asymmetric information on firms' abatement costs, the uncertainty of benefits from pollution control, incomplete enforcement, incomplete monitoring of emissions, the possibility of voluntary participation of non-affected sources and market power.

In Chapter 4, **John Reilly** and **Sergey Paltsev** examine the EU ETS by using a computable general equilibrium model of the world economy, disaggregated to provide detail on most of the major EU-25 countries and sectors covered in the ETS. They find that a competitive carbon market clears at a carbon price of ~0.6-0.9€/tCO_2 (~2-3€/tC) for the 2005-2007 period in line with many observers' expectations, but in sharp contrast to the actual early history of trading prices which had settled in the range of 20-25€/tCO_2 (~70 to 90€/tC) by the middle of 2005. They examine possible reasons for this divergence, from faulty modelling to firms' expectations about future periods and how that could impact prices in this period. Reilly and Paltsev argue that the performance of the ETS, real or perceived, could well affect the political acceptance of emissions trading as an instrument for managing the environment and, as such, that economists must carefully evaluate the maturing EU ETS.

European Union countries have agreed to reduce carbon dioxide emissions by individual amounts relative to their emissions in 1990 to reach the Kyoto target for the EU as a whole, in what is known as the Burden Sharing

Agreement (BSA). Given that the reduction requirements formulated within the BSA differ substantially among EU countries, in Chapter 5 **Tim Hoffmann, Andreas Löschel** and **Ulf Moslener** study the effects that this phenomenon may have on climate policy and compliance costs of EU Member States, particularly in the design of their National Allocation Plans to implement the EU ETS. They demonstrate that attempts to harmonize the free allocation of emission permits across countries may introduce significant efficiency costs, thus undermining the advantages of emissions trading, because it dramatically increases the variance of average costs per ton of reduced carbon dioxide within the sectors that are not subject to the ETS. The chapter shows that in some countries the average cost per ton of abated CO_2 increases substantially, so the use of the flexible mechanisms of the Kyoto Protocol (Joint Implementation and Clean Development Mechanism) is necessary to meet the target.

From 1990 to the end of 2005, Spanish carbon dioxide emissions grew by more than 45 percent, which is clearly incompatible with the EU allocation of Kyoto-mandated CO_2 reduction or BSA (+15 percent for Spain). The reasons for this situation are to be found in a combination of economic growth and an inefficient energy domain, coupled with a total absence of climate change policies. In Chapter 6, **Xavier Labandeira** and **Miguel Rodríguez** use a general equilibrium model to assess the effects of a sudden and intense (i.e., with a limited time to carry out the requested abatement) CO_2 reduction by the Spanish economy. The results, which could be valid for other developed Mediterranean economies, show that the costs of immediate and medium-size reductions are not significant in the short run and thus the system is a cost-effective way to attain environmental improvements. However, the authors confirm that delaying action with a binding commitment means that the degree of CO_2 emission reduction is much higher and so economic costs are far more important.

Pedro Linares, Francisco J. Santos and **Mariano Ventosa** start Chapter 7 by indicating that the use of renewable energies in the European Union has increased considerably in recent years, especially in those countries where price-based mechanisms have been used. However, there is also a discussion on whether this type of support is too generous for renewable producers, particularly in view of the EU ETS, which is expected to become an additional incentive for renewables through higher electricity prices. In this context, their objective is to analyze this impact quantitatively by simulating the behaviour of the Spanish electricity sector with and without the EU ETS, and specifically to look at the combined effect of the ETS and different renewable support systems in Spain. The chapter shows that, contrary to the expectations of some, the EU ETS will not induce significant changes in the profitability of renewable energies and therefore renewable installed power will not grow very much. However, the EU ETS may reduce the efficiency of the support systems already in place when these are price-based.

In Chapter 8, **Frauke Eckermann** uses a method for the verification of firm data within the EU ETS that sets the audit probability in relation to the allowance allocation and to penalties for misreporting. The chapter shows how, due to cost efficiency, a scheme with random audits and penalties should be preferred to an audit probability of one for each firm. In particular, the expected penalty should equal the gain from misreporting. Since penalties are limited in the political context, the author characterizes efficient verification schemes for different penalty structures. It is thus shown that in a scheme with linear penalties, the efficient audit probability is dependent on the form of the allocation function. For a regressive allocation of emission allowances, for example, the audit probability of an efficient verification mechanism is non-increasing in the reported values, while in the case of a proportional allocation of emission allowances, the audit probability should be the same for all firms. In a scheme with maximum penalties, the efficient audit probability is inversely related to the strength of the penalty and increasing in the reported emissions value for any allocation of emission allowances.

Following developments in the new growth theory, in Chapter 9 **Malte Schwoon** and **Richard S.J. Tol** start the third part of the book by presenting an analysis in which they include the costs associated with premature capital turnover by penalizing rapid changes in the level of abatement from one period to the next. The results suggest that the option to shift some abatement into the future, where it is less costly due mainly to discounting and also to learning, is more valuable than the option to increase current abatement so as to generate additional learning effects. Furthermore, the authors find that inertia is a much more important determinant of the optimal abatement path than induced technological change is.

In Chapter 10, **Rafaela Pérez** and **Jesús Ruiz**, in a second-best framework, analyze the dynamic properties of a general equilibrium model of endogenous growth with environmental externalities. In particular, they study the conditions for global and local indeterminacy and their implications for the control of pollution by governments. From a theoretical point of view, the existence of coordination failure between private agents and government can be explained through the indeterminacy of equilibria issue. As a consequence, the environmental policy implemented by government may not yield the effects on pollution levels that society wants. Therefore, the existence of this coordination failure can be explained through the indeterminacy of equilibria issue.

In Chapter 11, **Agustín Pérez-Barahona** and **Benteng Zou** study the effect of a tax on the energy expenditure of firms as a way to promote investments in energy-saving technologies. Fossil fuel is an essential input throughout all modern economies and thus the reduced availability of this basic input to production, e.g., due to climate change policies, would have a negative impact on GDP and economic growth through cutbacks in energy use. However, they show how this trade-off between energy reduction and growth could be less severe if energy conservation is raised by energy-saving technologies. To study such a hypothesis, the authors consider a general equilibrium model with embodied and exogenous energy-saving technological progress in a vintage capital framework, where the scrapping rule is endogenous and linear simplifications are eliminated.

In Chapter 12, **Carlos de Miguel, Baltasar Manzano** and **José M. Martín-Moreno** use a real business cycle model to analyze the consequences of oil price shocks on the characteristics of aggregate fluctuations and on welfare in a small open economy, as is the case of European economies. Their simulations indicate that oil shocks can account for a significant percentage of GDP fluctuations in many European countries, even though the explanatory power is quite smaller for others. In addition, the authors show that this model reproduces the cyclical path of the European economies in periods of oil crisis. Furthermore, it is shown that the increases in the relative price of oil have negative effects on welfare, which is particularly high in southern European countries.

Finally, in Chapter 13, **Adriaan van Zon** and **Tobias Kronenberg** present a cyclical growth model where the R&D sector is disaggregated, first into carbon and non-carbon R&D, and further into basic R&D and applied R&D. The authors argue that transitions from one energy system to the other can be better understood in a general purpose technology (GPT) framework because the energy sector relies on a small number of highly pervasive GPTs.

PART I

Some Fundamentals

2. Growth and environment: on U-curves without U-turns

Sjak Smulders

INTRODUCTION

Sooner or later any economist has to think on the following two questions that are in the mind of so many in our societies: 'How bad is economic growth for the environment?' and 'How bad is environmental policy for economic growth?' Detailed answers to these economy-wide questions are usually not be found in traditional textbooks on environmental economics because of their typical micro-economic focus. Nevertheless, almost all books provide an indication to the answers by showing one or two pictures.

The first is the pessimistic picture of the Club of Rome. Environmental quality declines over time because the economy has grown so much. We need a U-turn – a reduction in GDP – to arrive at a U-curve for environmental quality. The second picture is also a U-curve: the Environmental Kuznets Curve (EKC). Now GDP rather than time is on the horizontal axis. For some measures of environmental quality, it is empirically found that they first decline with income, but later improve. Once they have grown rich enough, countries start cleaning up.

The questions raised above are important enough to justify more than suggestive U-curves. They directly relate to climate change and energy issues. Can we expect climate change policy, and associated energy policies, to become more or less stringent when the world economy grows? Will technological change reduce the cost of climate change policy? Will a constant carbon tax mitigate global warming in a growing economy? Here I deal with these questions by examining the fundamental forces that shape the interaction between growth and the environment. I use graphical tools in the main text but the appendix provides the corresponding mathematical analysis.

THE ENVIRONMENTAL KUZNETZ CURVE (EKC)

There is considerable evidence that economic growth not necessarily deteriorates the environment. This evidence can be collected under the heading of the 'Environmental Kuznets Curve' (EKC) hypothesis. This literature tests how pollution changes with income levels. The relation between pollution and income is characterized as an EKC if pollution first rises with income, but starts to decline when income exceeds a certain threshold level. The hypothesis has been tested for several resources, energy sources and emissions. The evidence is mixed: the EKC shows up, but only for some pollutants and not in all countries (see Lieb, 2003 for a detailed overview of the results).

On theoretical grounds, there is no reason to expect pollution and growth to be unambiguously related, because both income and pollution are endogenous variables (cf. Copeland and Taylor, 2003). The pattern of growth, choice of technology and nature of technological change determine how income and pollution evolve over time and a host of underlying variables can affect both variables. Our aim can therefore only be to sort out the various basic forces that affect pollution and environmental degradation in the process of growth. We may expect an EKC if environmental policies sufficiently reflect social preferences for growth and environment, and thus reflect a rise in the demand for environmental quality. Alternatively, the EKC may arise without environmental regulation if technological change happens to be such that for low income the productivity of polluting inputs increases, but for higher income, it falls. This brief discussion reveals that, first, we need to carefully account for the institutional setting by making a distinction between optimal policies and (imperfectly regulated) market forces, and, second, that the main driving forces behind the pollution-growth link might come from preferences as well as technology.

THE ROLE OF PREFERENCES

We first explore how the Environmental Kuznets Curve (EKC) can be optimally driven by preferences for a clean environment, or – to be more precise – by satiation in demand for produced consumption goods. We do so reinterpreting and modifying the standard (textbook) diagram of allocating effort in a two-good economy, in which the optimal allocation is found by the point of tangency between iso-utility curves and the production frontier. For the more mathematically inclined reader, the appendix contains the mathematical formulation of the model and its main results.

The iso-utility curves in Figure 2.1 capture demand for environmental quality. Households care about produced consumption goods (*C*) and environmental quality (*N*). Hence we can draw the usual iso-utility functions *uu'* in a *C,N* plane. The better the two types of consumption are substitutable, the flatter the iso-utility curves are.

The straight lines in the figure represent production possibilities (the production frontier): they depict the feasible combinations of consumption and environmental quality. To concentrate on the preference side for the moment, we have simplified production structure as much as possible by assuming that one unit of the natural resource can be transformed into a given amount of consumption goods. This rules out substitution effects in supply and allows us to isolate the effect of substitution in demand. Lines further away from the origin represent more productive economies. Notice that a reduction in environmental quality by one unit can be interpreted as an increase in pollution by one unit. Hence, the linear production possibility curves assume that one unit of production always causes the same amount of pollution (or requires the same amount of polluting inputs).

Economic growth shifts out the production possibilities and both consumption and environmental quality increase in the optimum as shown in Figure 2.1. This simple example illustrates the power of demand for environmental quality: demand for both produced consumption and environmental quality goes up with income (since they are 'normal goods'), so that growth may produce higher environmental quality.

Environmental quality

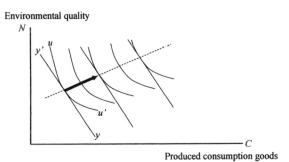

Figure 2.1   Normal goods

Our representation of production needs to be refined, however. A linear representation of production possibilities is inadequate without defining the end points of the line. First, man cannot do better than God: it is impossible to make earth cleaner than the paradise (the pristine level of environmental quality denoted by $\bar{N}$). Second, man needs the technical means to be as bad as the devil: the capacity to harvest the environment is limited by technology and production capacity (so the point where the linear production frontier cuts the horizontal axis is not necessarily attainable). Loggers in a poor village simply lack the technical means to cut all trees in the rain forest surrounding the village. Prehistoric man could hunt many deer, but lacked the capacity to destroy the ozone layer. Curve $\bar{N}\,y\,'y$ in Figure 2.2 represents the production possibilities under these restrictions: maximal environmental quality is at $\bar{N}$ is (the natural endowment or pristine level of the environment) and maximal production of consumption goods is at $C^z\ (K)$ if the production capacity is K. Note that this capacity constraint puts a lower bound on environmental quality as well. A higher stock of capital allows the economy to harvest more and to produce more, which is depicted as the outward parallel shifts of the $yy\,'$ lines. (Note that we retain our assumption of a (piecewise) linear transformation curve to abstract from substitution effects on the supply side and focus exclusively on substitution in demand.) Hence, if the capital stock (K) expands, points further to the right on the line $\bar{N}\bar{Z}$ become available.

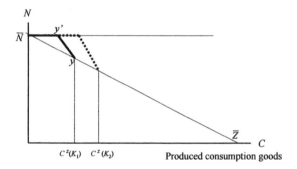

Figure 2.2 Not better than God, not worse than the devil

For a low production capacity level, the highest indifference curve is at the corner of maximal production (see Figure 2.3). It is the lack of capital that prevents the economy from substituting produced goods for environmental amenities up to the tangency point (where the rate of transformation exactly equals the rate of substitution). In this situation, environmental quality is not a scarce good and the capacity constraint is binding. An exogenous increase in the capital stock shifts out the production frontier along the $\bar{N}\bar{Z}$ line. Increases in production capacity are optimally used to increase consumption at the cost of the environment, until the production frontier has shifted out so much that its outer corner becomes tangent to the indifference curve. From this point on, environmental quality becomes a scarce good and the tangency point determines the optimal path. The result is a U-shaped relationship between environmental quality and income levels (EKC). Cleaning up eventually becomes optimal under very mild assumptions. The only thing we need is some '*satiation*': when produced consumption goods become more abundant, they are valued at a lower marginal utility relative to environmental amenities, that is the iso-utility curves become flatter for given N when C increases (Lieb, 2002).

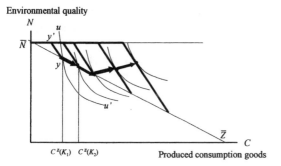

Figure 2.3   Too poor to pollute, too costly to abate

Not many models in the literature rely on this mechanism only. This is fortunate, since a concave production frontier, rather than the piecewise linear one in the figures above, seems more realistic. The present exposition just *isolates* the forces from the preference side that may drive the EKC. Almost all models in the static and growth literature (Selden and Song, 1995; Stokey, 1998; Lieb, 2002) incorporate the satiation effect, but combine it with a technology story similar to the one we will study in isolation in the next section.

THE ROLE OF TECHNOLOGY

To continue our strategy of isolating relevant effects, we now move from an extreme focus on preferences to an extreme focus on technology by assuming that produced consumption goods and environmental quality are perfect substitutes. This assumption rules out any degree of satiation as defined above; it implies that the indifference curves in the C, N plane are straight lines. Under these circumstances the preference side cannot explain the 'cleaning up phase'. Instead the EKC can emerge only because of changes in the opportunity cost of pollution.

In Figure 2.4 the curve $\bar{N}y$ depicts production possibilities (which are no longer piecewise linear as above). In the present case, the cost at which environmental amenities can be exchanged for produced goods varies with size and allocation of production capacity. Point y on the $\bar{N}\bar{Z}$ line corresponds to the situation that the full capital stock is allocated to the production of produced consumption goods (C). Departing from this point, environmental quality can be improved by allocating some capacity to specific abatement activities, which come at the cost of (net) production of produced consumption goods. This is represented by moving North-West along the curve $\bar{N}y$. Initially, the cost of abatement in terms of foregone consumption is small, as reflected by the steep slope, but, when approaching maximum environmental quality $\bar{N}$, costs rise, as is reflected by the flatter slope. Thus, the concave curvature of $\bar{N}y$ reflects that abatement costs in terms of foregone production C increase with environmental quality N.

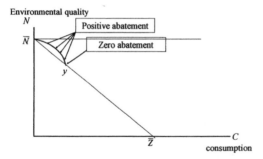

Figure 2.4   Consumption versus abatement

When production capacity (the capital stock) expands, the production possibilities curve shifts out. Production without abatement (with capacity allocated to production only) remains located at line $\bar{N}\bar{Z}$, which we call the zero-abatement line. Figure 2.5 assumes that doubling both capacity and abatement efforts doubles human impact on the environment. With our linear indifference curves, optimal environmental quality falls. Hence, we find that with constant returns to scale, in a growing economy, pollution increases and environmental quality declines by the technology effect. The basic reason is that higher production capacity makes exploitation of the environment more attractive. At a given level of environmental quality, the cost of improving it rises with production capacity.

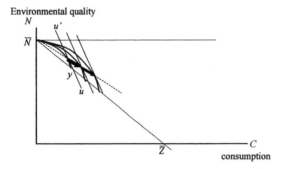

Figure 2.5   Factor accumulation

Technical progress that improves factor productivity has similar effects as factor accumulation. This allows for more production without necessarily increasing prollution. The production frontier shifts out as in Figure 2.6 from y to y' (the zero-abatement ($\bar{N}\bar{Z}$) line, not drawn, becomes flatter). Hence we get the perhaps surprising result that technical progress which allows producing the same amount with less pollution results in a higher optimum level of pollution. The reason is that technical change in this situation again increases in the opportunity cost of environmental improvements. A given increase in environmental quality leads to a bigger reduction in production (as reflected by the flatter slope of the production possibilities curve), since technological change makes polluting inputs more productive.

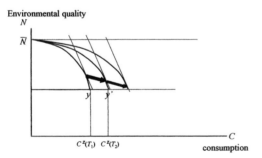

Figure 2.6   Cleaner technology

When does technological change trigger cleaning up rather than more pollution? For this we need improvements in the trade-off between production and pollution for given levels of inputs. For instance, new technologies allow given reductions in pollution at lower costs. This can be best interpreted as improvements in *abatement technology*, as opposed to improvements in production technology. Figure 2.7 shows this by shifting out the production frontier in the region where there is some abatement.

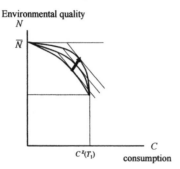

Figure 2.7   Improved abatement technology

The EKC emerges if initially production technology improves, but later on abatement technology improves. This happens if the average cost to abate pollution (the cost to improve the environment in terms of reduced net output) decreases with the level of abatement, because of economies of scale or learning by doing (cf. Andreoni and Levinson, 2001). In a growing economy, the production possibility curves then not only shift out, but also become steeper, as is drawn in Figure 2.8. When we confront this with our linear indifference curves, we see the EKC emerge. The basic reason is that: with a larger scale of economic activity, pollution becomes more and more productive, but also abatement becomes less and less costly. In a very large economy, it costs basically nothing to raise N to $\bar{N}$.

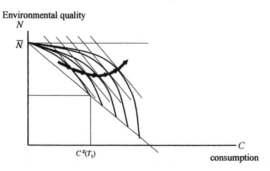

Figure 2.8   Big enough to abate cheaply

MARKET INCENTIVES AND IMPERFECT REGULATION

Up to now we studied the social optimum. What happens in a decentralized economy in which market forces together with certain policies determine allocation? In fact, anything can happen, depending on how policies are set. At one extreme, the optimal policy ensures that polluters fully internalize social costs and benefits of changes in environmental quality. At the other extreme is a *laissez-faire* economy with no price of pollution. Many intermediate cases of suboptimal pollution policies can be conceived. A useful benchmark is the case of a constant pollution tax.

For any equilibrium point in our diagrams, we can find the corresponding prices in the (regulated) market, in particular the price of pollution (or pollution tax). With perfect competition and no other externalities than pollution externalities, firms equate the marginal product of pollution to the pollution tax. In our simple setting, the slope of the production possibilities curve in the equilibrium point is equal to the inverse of marginal product of pollution, since this slope gives the change in environmental quality (which equals the change in pollution) needed to marginally change goods production, C. Hence, we can infer from our diagrams how the pollution tax changes over time by looking at the slope of the production possibilities curve.

The optimal tax can be derived from the above figures. In Figure 2.3, the *optimal tax* is zero in the first phase (when corner solutions apply and environmental quality falls), but becomes positive in the second stage (when environmental quality rises); the pollution tax is constant, but this follows directly from the assumption of perfect substitution in production. In Figure 2.5, Figure 2.6, Figure 2.7, and Figure 2.8, we have assumed perfect substitutes between environmental quality and production, which amount to a constant social price of pollution in terms of consumption goods; hence the optimal pollution tax is constant in these cases as well. In the more general case with imperfect substitution in both production and utility, and constant returns to scale (or close enough to that), the optimal tax rate increases over time. (See appendix for more precise results.)

If the pollution tax rate is *zero*, there is no role for environmental preferences since environmental effects are a pure externality not internalized by any policy. There will be no abatement (by the definition of abatement). The equilibrium is the point on the production possibilities frontier that is tangent (corner) to a vertical line. Note that a zero tax does not produce infinite amounts of pollution since there is a maximum capacity to pollute as reflected by $C^z(K)$.

We finally consider the case of an *arbitrary constant tax rate*, which is in principle suboptimal. The question is how pollution evolves over time if the pollution tax does not change over time. Accumulation of capital and total factor productivity growth raise the marginal product of polluting inputs, so they will raise pollution given the constant pollution tax firms face. Technological progress that makes it easier to substitute away polluting inputs (that is technological progress in abatement) and cheaper abatement makes firms pollute less. Pollution will only fall if reductions in abatement costs are large enough to outweigh the effects of capital accumulation.

When does the EKC emerge in a market economy without optimal (Pigouvian) taxation? Of course, in a second best world a lot can happen, depending on the various externalities. The simplest story is one in which taxes are initially low or even zero, but rise in the advent of damage and policy action. An EKC emerges in this case. If exogenous improvements in abatement technology are slow initially, but fast enough later on, the EKC emerges even with a constant tax.

CONCLUSIONS

The analysis has sorted out three basic mechanisms driving the relationship between income and pollution or environmental quality: satiation in demand, expansion of production after capacity and technological improvements. With growing production of produced goods, environmental quality becomes relatively scarce and demand for it is likely to increase, thus favouring improvements in environmental quality, provided proper policies translate society's demand into incentives and action. Larger production capacity raises the cost of environmental policy, so that pollution may grow with income. Finally, technological change may lower the cost of abatement and might overcome the effect of larger production capacities. Other types of technological change are likely to make the economy more productive and therefore are likely to also make polluting inputs more productive; hence technological change in many cases increases the incentive to pollute more. Growth driven by technological change increases pollution on this account.

The three effects – demand, capital accumulation and technological change – operate simultaneously. Therefore there is no reason to expect the EKC to emerge even if environmental policies are internalizing all externalities. It is important to note that for the case that optimal pollution follows the EKC pattern, there exists a path of the tax rate for which pollution in the regulated market economy does not follow an EKC pattern, and *vice versa*. As was shown, it is ambiguous whether optimal pollution follows the EKC. Similarly, there is no reason to expect an EKC in equilibrium. All these results are consistent with empirical evidence, which shows that the EKC only emerges in a limited number of cases.

MATHEMATICAL APPENDIX

Optimal environmental policy

The mathematical formulation of the model to study optimal environmental policy can be written as the following maximization problem:

$$\max_P \ U(\overset{+}{C},\overset{+}{N}) \tag{1}$$

$$\text{s.t. } C = T \cdot F(\overset{+}{K},\overset{+}{P}) \tag{2}$$

$$N = N(\overset{-}{P}) \tag{3}$$

$$P \leq \overset{+}{P^z}(K) \tag{4}$$

Here C, N, K, and P denote consumption, environmental quality, man-made capital and pollution, respectively. To establish the connection to the figures in the main text, we define maximum consumption as $C^z(K) = T \cdot F(K, P^z(K))$ and assume $N_P = -1$.

The plus and minus signs denote the signs of the first order derivatives. K and T are exogenous in this static model. T captures TFP or clean factor growth. We put a little bit more structure on the model to derive clear results. In particular, we make the convenient assumption that the functions U and F are homogenous of degree one and can be characterized by elasticities of substitution σ_{CN} and σ_{KP} respectively (which are possibly non-constant).

The optimality condition reads:

$$\left(\frac{U_c}{U_N}\right)^{(>)}_{=}\left(\frac{-N_P}{Y_P}\right) \text{ and } P \overset{(=)}{<} P^z(K) \tag{5}$$

where subscripts denote partial derivatives. We discuss in turn the interior and corner solutions.

Interior solution: does the cleaning-up phase exist?
Differentiating condition (5) (with equality and inequality sign respectively) reveals how optimum pollution changes with accumulation of K and with TFP growth:

$$\hat{P} = \frac{1}{\Delta_1}\left[\left(\frac{1}{\sigma_{PK}} - \frac{1}{\sigma_{CN}}\right)\theta_K \hat{K} + \left(1 - \frac{1}{\sigma_{CN}}\right)\hat{T}\right] \qquad (6)$$

Here, hats denote growth rates, Δ_1 is a positive number, σ_{PK} (σ_{CN}) is the elasticity of substitution in production (utility), and θ_K is the production elasticity of K. We draw the following conclusions.

- *What matters is substitution (in F and U).* Factor accumulation raises (lowers) pollution if substitution in utility is easier (more difficult) than substitution in production. A low elasticity in production implies that it is costly to give up pollution, while a high elasticity in utility implies that it is not very costly to exchange some produced goods for environmental quality.
- *The sources of growth matter.* Capital accumulation and technological progress have different effects. TFP growth lowers pollution only if substitution in utility is poor. With high substitution in utility, it is optimal to exploit the higher productivity of polluting inputs to raise production.
- In Figure 2.1, we assume $\sigma_{PK} \rightarrow \infty$ and $\hat{T} = 0$ so that $\hat{P} / \hat{K} < 0$.
- In Figures 2.5–8, we assume $\sigma_{CN} \rightarrow \infty$ so that $\hat{P} / \hat{K} > 0$ and $\hat{P} / \hat{T} > 0$ (Figure 2.5 and 2.6, respectively).
- In the main text we argue that pollution-saving technological change amounts to a more curved production possibilities frontier. In the mathematical model this corresponds to a larger elasticity between polluting inputs and capital in the net production function. We assume that the production function in (2) is of the CES specification, i.e.:

$$F(K,P) = \left[\alpha K^\rho + (1-\alpha)P^\rho\right]^{1/\rho}$$

so that $\sigma_{KP} = 1 / (1 - \rho)$ is a parameter. Moreover, we specify the maximum-pollution function in (4) as $P^z(K) = K$. Allowing for an exogenous change in σ_{PK}, we find instead of (6):

$$\hat{P} = \frac{1}{\Delta_1}\left[\left(\frac{1}{\sigma_{PK}} - \frac{1}{\sigma_{CN}}\right)\theta_K \hat{K} + \left(1 - \frac{1}{\sigma_{CN}}\right)(\hat{T} - \eta_{Yp\sigma}\hat{\sigma}_{KP})\right] \qquad (7)$$

where

$$\eta_{YP\sigma} = \left((\sigma_{KP} - 1)\ln\frac{1}{1-\theta_K} + (1-\theta_K)\ln\frac{1}{1-\theta_K} + \theta_K\ln\frac{1}{\theta_K} \right)\frac{1}{(\sigma_{KP}-1)^2}$$

- In Figure 2.7, we assume $\sigma_{CN} \to \infty$ so that $\hat{P}/\hat{\sigma}_{PK} < 0$, provided $\eta_{YP\sigma} > 0$, which requires $\sigma_{KP} > 1$ or N not too close to $\bar{N}$. Then, better substitution implies lower pollution. We see that this result more enerally arises whenever $\sigma_{CN} > 1$ and $\eta_{YP\sigma} > 0$.

- The Androni Levinson (2001) result, visualized in Figure 2.8, shows up as follows. Suppose, initially – that is for a small level of K – we have $\sigma_{PK} < \sigma_{CN}$, so that pollution rises with the capital stock, $\hat{P}/\hat{K} > 0$. In the main text we have shown that increasing returns in abatement technology amounts to an outward-shifting and more curved production possibilities frontier when K increases. To capture this we assume that σ_{PK} rises with K. Hence, after sufficient accumulation of K, the inequality will be reversed, $\sigma_{PK} > \sigma_{CN}$, and pollution falls.

Remarks
Technological progress may take many other forms: we abstracted from factor-augmenting technological progress and technological progress that affects (for given levels of K) the zero-abatement pollution level P^z.

The CES function is not a very general specification for utility or production (but more general, of course, than Cobb Douglas). The equation above only holds as a local result for more general functions. Even within the class of CES functions, the result above is not general. With more than two factors or with technical progress that is not of the TFP form, substitution cannot any longer be represented by a single elasticity of substitution and the nesting becomes important.

An alternative specification for utility proves very useful. Suppose U is additively separable in C and N, but we maintain diminishing marginal utility with respect to both arguments. It turns out that the expression above should be modified to:

$$\hat{P} = \frac{1}{\Delta_2}\left[\left(\frac{1}{\sigma_{PK}} - \eta_{Ucc} \right)\theta_K\hat{K} + (1-\eta_{Ucc})(\hat{T} - \eta_{Yp\sigma}\hat{\sigma}_{KP}) \right]$$

where η_{Ucc} is the elasticity of marginal utility of consumption, positively defined (see Lopez, 1994; the additively separable utility function is frequently used in the literature, e.g. Stokey, 1998; Aghion and Howitt, 1998). If marginal utility falls quickly with C, i.e. if $\eta\ U_{cc}$ is large, giving up some consumption in exchange for environmental quality is not very costly and pollution tends to fall with capital accumulation or TFP growth.

Corner solution: does technological progress help in the growing dirty phase?
We have seen that corner solutions cannot be ruled out, see (5). If the marginal product of pollution is large at zero abatement levels, pollution will be at its maximum P^z, which rises with K. It has to be non-declining with any kind of technological progress too. Suppose technological changes shifts the P^z function (instead of (4), write $P \le P^z\ (T_p^+ K)$). Technological progress is a case in which the same amount of capital allows for higher levels of consumption. This happens if P^z rises for given levels of K (i.e. T_P rises). It is as if the economy can exploit more natural resources. It is clear that this leads to more pollution in a corner solution.

Pollution tax

Firms equate the marginal product of pollution Y_P to the tax, denoted by τ. First consider the *socially optimum pollution tax*. The optimality condition (5) tells us that the tax in the social optimum is the Pigouvian tax $\tau{=}Y_P{=}{-}N_P U_N/U_C$. The tax equals marginal damage, which can be thought of as consisting of two components: first, the psychological marginal damage (U_C/U_N) and, second, the ecological marginal damage $-N_P$. Hence, we can write the change in the the optimal tax rate, τ^*, as:

$$\hat{\tau}^* = \left(\frac{1}{\sigma_{CN}}\right)\hat{C} + \left(\frac{\eta_{NP}}{\sigma_{CN}} + \eta_{NPP}\right)\hat{P}$$

where the η parameters are positively defined elasticities. The expression simply equates the change in the tax to the change in marginal damage. Since all elasticities are positive, the pollution tax must rise when the economy is growing and pollution is non-decreasing. If pollution is decreasing, ecological marginal damage falls $(\eta_{NPP}{>}0)$, which allows for a slow down in tax rises. Moreover, psychological marginal damage falls $(\eta_{NP}/\sigma_{CN}{>}0)$, which also allows for lower taxes.

Next consider the case of an *arbitrary tax rate*. The question is how pollution evolves over time. We find this by evaluating the growth of Y_P and setting it equal to the change in the tax:

$$\hat{P} = \hat{K} + \frac{\sigma_{KP}}{\theta_K}\hat{T} - \frac{\sigma_{KP}}{\theta_K}\hat{\tau} - \eta_{YP\sigma}\hat{\sigma}_{KP}$$

Accumulation of polluting capital and total factor productivity growth improve the marginal product of polluting inputs, so they will raise pollution. Rising pollution taxes decrease pollution. Technological progress that makes it easier to substitute away polluting inputs (that is technological progress in abatement) is captured by a rise in σ_{KP}. Since $\eta_{YP\sigma}$ is most likely a positive number, cheaper abatement makes firms pollute less. However, if capital is an important input ($\sigma_{PK}<1$ and θ_K large so that $\eta_{YP\sigma}<0$), it is more attractive to raise the productivity of K by increasing P. Since capital is so productive, doing so generates enough revenue to both pay the higher pollution tax bill and sell more.

REFERENCES

Aghion, P. and P. Howitt (1998), *Endogenous growth theory*, Cambridge: MIT Press.

Andreoni, J. and A. Levinson (2001), 'The simple analytics of the Environmental Kuznets Curve', *Journal of Public Economics*, **80**, 269-286.

Copeland, B. and S. Taylor (2003), *International Trade and the Environment: Theory and Evidence*, Princeton: Princeton University Press.

Lieb, C.M. (2002), 'The environmental Kuznets curve and satiation: a simple static model', *Environment and Development Economics*, 7, 429-448.

Lieb, C.M. (2003), 'The environmental Kuznets Curve, a survey of the empirical evidence and of possible causes', Heidelberg, department of economics, discussion paper series, 391.

Lopez, R. (1994), 'The environment as a factor of production: the effects of economic growth and trade liberalization', *Journal of Environmental Economics and Management*, **27**, 163-184.

Selden, T. and D. Song (1994), 'Neoclassical growth, the J curve for abatement and the inverted U curve for Pollution', *Journal of Environmental Economics and Management*, **29**, 162-168.

Stokey, N.L. (1998), 'Are There Limits to Growth?', *International Economic Review*, **39**, 1-31.

3. Pollution markets: some theory and evidence

Juan P. Montero

INTRODUCTION

Policy makers in different parts of the world are paying more attention to pollution markets (i.e., tradeable emission permits markets) as an alternative to the traditional command-and-command control (CAC) approach of setting uniform emission and technology standards. A notable example is the 1990 US Acid Rain Program that implemented a nationwide market for electric utilities' sulfur dioxide (SO_2) emissions (Schmalensee *et al.*, 1998; Ellerman *et al.*, 2000). The US Environmental Protection Agency's (EPA) emissions trading policy represents another and older attempt to implement permits markets to mitigate air pollution problems in urban areas across the country (Hahn, 1989; Foster and Hahn, 1995; Tietenberg, 1985). In addition, a few less developed countries are also beginning to experiment in different forms with emissions trading (Montero *et al.*, 2002; Stavins, 2003; Tietenberg, 2005).

These experiences should not leave the impression that pollution permit markets have come anywhere close to replacing the traditional command-and-control approach. More reason to believe that permit markets are expected to play an increasing role in the solution of environmental problems in the future. In this sense, experience with existing permit markets help understand the importance of practical considerations for the design and implementation of these markets and for establishing the conditions under which they are likely to perform better than alternative instruments. My intention in this chapter is not to provide an exhaustive treatment of all practical considerations that may prove relevant, but only some of those that have caught my attention as I review the performance of existing permits programs (particularly the Acid Rain Program in the US and the total suspended particulate program of Santiago), and proposals for implementation of new ones (particularly carbon trading for dealing with global warming and a comprehensive permits program for curbing air pollution in Santiago).

Motivated by the operation these programs, in the rest of this chapter I extend the basic (perfect information, perfect competition) model of a permits market to accommodate for several practical considerations: regulator's asymmetric information on firms' abatement costs, uncertainty on benefits from pollution control, incomplete enforcement, incomplete monitoring of emissions, the possibility of voluntary participation of non-affected sources and market power. Implications for instrument design and implementation are provided. At the end I discuss topics for further research.

BASIC MODEL

Consider a continuum of firms of mass 1. In the absence of environmental regulation, each firm emits one unit of pollution which can be abated at a cost c. The value of c, which is private information, differs across firms according to the (continuous) density function $g(c)$ and cumulative density function $G(c)$ defined over the interval $[c_l, c_h]$. These functions are known by the welfare-maximizing regulator. Although the regulator does not know the control cost of any particular firm, he can derive the aggregate abatement cost curve for the industry, $C(q)$, where $0 \leq q \leq 1$ is the aggregate quantity of emissions reduction. The regulator also knows that the benefit curve from emissions reduction in any given period is $B(q)$. As usual, I assume that $B'(q)>0$, $B''(q)\leq 0$, $C'(q)>0$, $C''(q)\geq 0$, $B'(0)>C'(0)$, and $B'(q)<C'(q)$ for q sufficiently large.

Letting the regulator's welfare function be $W(q)=B(q)-C(q)$, the first-best reduction level q^* solves $B'(q^*)=C'(q^*) = c^*$, where $G(c^*)=q^*$. It is first-best optimal that firms with costs equal or below c^* be the only ones reducing emissions. To implement the first-best solution the regulator can either set a Pigouvian tax on emissions equal to $\tau = c^*$ or allocate a total of $x=1-q^*$ tradeable emission permits (recall that aggregate counterfactual emissions are equal to 1). If the regulator introduces a tax τ, firms with $c<\tau$ will reduce emissions while firms with $c>\tau$ will prefer to pay the tax. Thus, when τ is set at the first-best level c^*, firms will have incentives to reduce exactly up to the first-best level q^*.

If, on the other hand, the regulator distributes x permits either for free or through an auction, the market clearing price will be $p=C'(1-x)=G^{-1}(1-x)$. In particular, if the regulator allocates to each firm x permits for free, firms with $c>p$ will be making no reductions and buying extra permits to cover their emissions while firms with $c<p$ will be reducing their emissions and selling all

their permits. Thus, when x is set at the first-best level $1-q^*$, the resulting clearing price will be exactly at the first-best level c^*.

In this particular setting in which the regulator knows both the aggregate abatement cost curve and the benefit curve, he is clearly indifferent as to whether to use a price instrument (taxes) or a quantity instrument (permits) to reach the first-best solution. More generally, he can use either taxes or permits to achieve any emission goal (other than $1-q^*$) at the lowest cost.

In practice, however, we rarely see regulators using taxes or permits. With a few exceptions, they almost exclusively rely on the traditional CAC approach of setting (uniform) emission and technology standards. Under this approach, a regulator with an aggregate emission goal of x would require each firm to emit no more than x. Clearly, this approach will result in an inefficient allocation of abatement across firms unless they have identical abatement costs (i.e., $c_l=c_h$), which is unlikely. As typically occurs under standards, high cost firms are reducing too much while low cost firms are reducing too little. Because of this efficiency loss, economists have been long arguing for the wider use of market-based instruments such as permits (Montgomery, 1972).

Leaving aside political economy considerations that may help explain the limited use of market-based instruments (Stavins, 2003) in the remainder of this chapter I will extend this basic model to incorporate additional elements that regulators are likely to face in the practical design and implementation of environmental markets.

EXTENDING THE BASIC MODEL

In extending the basic model, it is important to keep in mind that the normative implications of these practical considerations may affect the policy design in various ways that can go from a simple tightening of the basic design, to a combination of permits with some other instrument (such as taxes or standards), or yet, to the replacement of permits by an alternative instrument.

Imperfect information on costs and benefits

Several authors have extended the basic model to the case in which the regulator knows little about firms' costs but can costlessly monitor each firm's actual emissions and enforce compliance. To capture the regulator's imperfect information on costs in our model, let his prior belief be $c(\theta)=c+\theta$, where θ is some stochastic term such that $E[\theta]=0$ and $E[\theta^2]>0$, where $E[.]$ is the expected value operator. I assume that θ is common to all individual costs $c \in [c_l, c_h]$, which produces the desired 'parallel' shift of the aggregate marginal cost curve, $C'(q)$, by the amount θ. In other words, $C'(q,\theta)=C'(q)+\theta$. Recall that $c(\theta)$ continues to be firm's private information, so the realization of θ is known by all firms before they make and implement their compliance (and production) plans.

Because the introduction of θ leaves the regulator uncertain about the true aggregate marginal cost curve, he can no longer implement the first-best solution by simply allocating a certain number of permits. Making use of the revelation principle, Kwerel (1977) and Dasgupta *et al.* (1980) show that this information asymmetry does not prevent the regulator from achieving the first-best if he can use non-linear instruments (i.e., transfer to or from firms contingent on their cost revelations and emissions).[1]

Despite the welfare superiority of these non-linear instruments, experience shows that regulators always favor simple regulatory designs that can be implemented in practice. For this particular reason it remains relevant to understand the implications of imperfect information on the design of relatively simple instruments such as permits, (linear) taxes and standards.

While cost uncertainty does not change the welfare advantage of permits over standards, in a seminal paper Weitzman (1974) showed that it does break the welfare equivalence between permits and taxes. To expand the basic model in a tractable way let us follow Weitzman (1974) and Baumol and Oates (1988) and consider linear approximations for the marginal benefit and cost curves and additive uncertainty. We will also allow the regulator be uncertain about benefits, so $B'(q,\eta)=B'(q)+\eta$, where η is a stochastic term such that $E[\eta]=0$ and $E[\eta^2]>0$.

It is not difficult to show that under the linearity assumptions the optimal tax and permits design are as in the certainty case, that is $\tau=c^*$ and $x=1-q^*$. Because of uncertainty, however, neither design will be optimal *ex-post* (unless $\eta=\theta=0$). The relevant policy question then is which instrument is expected to come closer to the *ex-post* optimum. To explore this question we estimate the difference between the expected social welfare under the price instrument (taxes) and that under the quantity instrument (permits). Solving (assume $E[\eta\theta]=0$) we obtain the famous Weitzman's (1974) formula:

$$\Delta_{pq} = \frac{E[\theta^2]}{2(C'')^2}(B''+C'')$$ (1)

where Δ_{pq} is the expected welfare difference between using the price instrument and the quantity instrument, $B''<0$ is the slope of the (linear) marginal benefit curve and $C''<0$ is the slope of the (linear) aggregate marginal cost curve.

Part I: Some Fundamentals

The normative implications of (1) are quite clear: prices (i.e., taxes) ought to be preferred if the marginal cost curve is steeper than the marginal benefit curve; that is, if $C''>|B''|$; otherwise quantities (i.e., permits) ought to be preferred. The rationale for using prices over quantities is the following. As long as miscalculating the *ex-post* optimum amount of control has lower welfare consequences than miscalculating the *ex-post* optimum (marginal) cost of control, which happens when the marginal cost curve is steeper than the marginal benefit curve, prices are preferred.

In a quantity regime the amount of control remains fixed while the marginal cost of control (i.e., permits price) is uncertain to the regulator. Conversely, in a price regime the marginal cost of control remains fixed (equal to the tax level) while the amount of control is uncertain. So, from a regulator's perspective the quantity regime is superior to the price regime in terms of (higher) expected benefits from pollution control but the price regime is superior in terms of (lower) expected costs. Then, for example, if the marginal cost curve is very steep, the (marginal) cost of control can deviate significantly from the *ex-post* optimum; situation in which a price instrument that fixes the marginal cost of control turns more appropriate. Note that benefit uncertainty is absent unless there is some correlation between η and θ.

Because neither permits nor taxes are *ex-post* optimum, there seems to be room for a hybrid policy to improve upon either single-instrument policy. Roberts and Spence (1976) formally showed that a hybrid policy that combines $x=1-q^*$ permits with a tax $\tau>c^*$ and subsidy $s<c^*$ is always superior to either single-instrument policy.[2] If, for example costs happen to be higher than expected, i.e., $\theta>0$, the allocation of $1-q^*$ permits appear too tight *ex-post* resulting in too high prices. The introduction of the tax puts a ceiling on the permits price, which is equivalent as to having the regulator issuing additional permits. If, on the other hand, costs happen to be lower than expected, i.e., $\theta<0$, the allocation of $1-q^*$ permits appear too lenient *ex-post* resulting in too low prices. The introduction of the subsidy puts a floor on permits prices, which is equivalent as to having the regulator buying-back some permits.[3]

The idea of combining permits with taxes (but less with subsidies) is at the center of the debate on instrument design for reducing carbon dioxide emissions believed to be responsible for global warming. Early proposals had permits as the only single instrument to reduce these emissions (see, e.g., Kyoto Protocol), but because several studies have shown compliance costs to be quite uncertain, more recent proposals argue for the inclusion of a tax as a safety valve in case the price of permits climbs inefficiently high (Pizer, 2002).

Incomplete enforcement

It is well known that regulations are not always fully enforced; the TSP program in Santiago is a good example of that. To understand the implications of incomplete enforcement on policy design, in Montero (2002a) I extend the Weitzman (1974) analysis. The regulator is responsible for ensuring the individual firms' compliance with either the price or the quantity instrument. Firms are required to monitor their own emissions and submit a compliance status report to the regulator. Emissions are not observed by the regulator except during costly inspection visits, when they can be measured accurately. Thus, some firms may have an incentive to report themselves as being in compliance when, in reality, they are not.

The cost of each inspection is assumed to be large enough that full compliance is not socially optimal (Becker, 1968).[4] Therefore, in order to verify reports' truthfulness, the regulator randomly inspects those firms reporting compliance through pollution reduction to monitor their emissions (or check their abatement equipment). Each firm reporting compliance faces a probability ϕ of being inspected. Firms found to be in disagreement with their reports are levied a fine F ($\leq$F^{max}, where F^{max} is the maximum feasible fine, which value is beyond the control of the regulator) and brought under compliance in the next period.[5] To come under compliance, firms can reduce pollution or, depending on the regulatory regime, either pay taxes or buy permits. Firms reporting noncompliance face the same treatment, so it is always in a firm's best economic interests to report compliance, even if that is not the case.[6] Finally, I assume that the regulator does not alter its policy of random inspections in response to information acquired about firms' type, so each firm submitting a compliance report faces a constant probability ϕ of being inspected.

After deriving the optimal price and quantity design under uncertainty and incomplete enforcement (designs that include, among others, F=F^{max}), the Weitzman comparison between prices (i.e., taxes) and quantities (i.e., permits) shown in (1) changes to:

$$\Delta_{pq} = \frac{\gamma E[\theta^2]}{2(C'')^2}\left((2-\gamma)B''+C''\right) \tag{2}$$

where $\gamma=\phi/(1+\phi)<1$ is the fraction of the non-compliant firms (i.e., all those firms that have incentives to submit false reports) that are in compliance in any given period. Since $2-\gamma >1$, eq. (2) shows that incomplete enforcement improves the relative advantage of permits over taxes.

To explain this result requires first to understand that the presence of incomplete enforcement makes the effective (or observed) amount of control under a quantity instrument no longer fixed, as in a permits program with full compliance. Instead, it adapts to the actual cost of control. Indeed, if the marginal cost curve proves to be higher than expected by the regulator, more firms would choose not to comply, and consequently, both the effective amount of control and the cost of control would be lower than expected.

The fact that the effective reduction now becomes uncertain has two effects on the welfare comparison between prices and quantities that can be explained expanding (2) to:

$$\Delta_{pq} = \gamma(2-\gamma)\frac{E[\theta^2]B''}{2(C'')^2} + \gamma\frac{E[\theta^2]C''}{2(C'')^2} \tag{3}$$

where the first term is the difference in expected benefits, which is negative unless $B''=0$, and the second term is the difference in expected costs, which is always positive. Now, the first effect of incomplete enforcement on the welfare comparison is captured in the first-term of the right hand side, which shows that the advantage of quantities over prices on the benefit side is reduced to $\gamma(2-\gamma)<1$ relative to the case of full compliance (i.e., $\gamma=1$). The second effect is captured in the second-term of the right hand side of (3) that shows that the advantage of prices over quantities on the cost side is also reduced to $\gamma<1$. Because $\gamma(2-\gamma)>\gamma$, the second effect dominates and the overall advantage of prices over quantities is reduced. From (3) one also argue that incomplete enforcement makes both the marginal benefit curve and the marginal cost curve to look flatter, but because $\gamma(2-\gamma)>\gamma$, it makes the marginal cost curve even more so. In addition, note that as γ falls, the welfare difference between the two instrument shrinks and disappears when there is no compliance at all (i.e., $\gamma=0$).

Another way to interpret this result is that incomplete enforcement softens the quantity regime, making it resemble a non-linear instrument, as in Roberts and Spence (1976). Indeed, when costs prove to be higher than expected, some firms choose not to comply, increasing the effective amount of pollution.

Multipollutant markets

In dealing with Santiago air pollution, or more generally, in any urban pollution control effort, the design and implementation of good environmental policy necessarily involves more than one pollutant or different zone (e.g., inland and coastal zones). Hence, the study of permit programs to simultaneously regulate various pollutants becomes very relevant. If the regulator has perfect information about costs and benefits of pollution control for each of the pollutants involved, it is evident that the regulator can implement the first-best through the allocation of permits to the different markets without the need for interpollutant trading. Under imperfect information on costs and benefits and possibly partial compliance, in Montero (2001) I show that the optimal permits design is more involved. It may be (second-best) optimal, under some conditions, to have the different pollutant markets integrated through some optimal exchange rates. In practical terms, it may be optimal allowing firms to cover their emissions of particulate matter (PM10) with permits of nitrogen oxides (NOx). Obviously some exchange rate must be defined.

To study under what conditions market integration is beneficial, I use the Weitzman framework and compare welfare under market integration vs. welfare under market separation. I consider two pollutants 1 and 2 (e.g., PM10 and NOx). If I impose some symmetry to the problem, that is $B''_1=B''_2=B''$, $C''_1=C''_2=C''$, $\phi_1=\phi_2=\phi$ and θ_1 and θ_2 are i.i.d. and not correlated with η (the intercepts of the marginal curves and the F's can vary across markets), the optimal amount of permits to be distributed under integration is the same that under separation. In addition, it is possible to establish that the welfare advantage of having the two markets working together (t) over separately (s) is given by the (familiar) expression:

$$\Delta_{pq} = \frac{\gamma E[\theta^2]}{2(C'')^2}\left((2-\gamma)B''+C''\right) \tag{4}$$

where $\gamma=\phi/(1+\phi)<1$ is, as before, the fraction of the non-compliant firms that are in compliance in any given period. Recall that $E[\theta^2]$ captures regulator's uncertainty about firms' costs, $B''<0$ is the slope of the (linear) marginal benefit curves and $C''>0$ is the slope of the (linear) marginal cost curves.

The first in eq. (4) is that under full enforcement $\gamma=1$) the regulator should allow interpollutant trading (i.e., market integration) as long as the marginal cost curves are steeper than the marginal benefit curves. This result is analogous to the result obtained by Weitzman (1974), a similar rationale applies to our multipollutant markets story. Interpollutant trading provides more flexibility to firms in case costs are higher than expected, but at the same time, it makes the amount of control in each market more uncertain. Then, if the marginal cost curves are steeper than the marginal benefit curves, the regulator should pay more attention to cost of control rather than the amount of control, and therefore, have markets integrated. On the other hand, if the marginal benefit curves are steeper than the marginal cost curves, the regulator should pay more attention to the amount of control in each market, and therefore, have markets separated. The presence of incomplete enforcement ($\gamma < 1$) has important effects on the multipollutant markets design as well. Since $2-\gamma > 1$, (4) indicates that incomplete enforcement reduces the advantage of market integration: the regulator should allow interpollutant trading only if the marginal cost curves are $2-\gamma$ times steeper than the marginal benefit curves.

Incomplete monitoring

Most market experiences implemented so far suggest that conventional permits programs are likely to be used in cases where emissions can be closely monitored, which almost exclusively occurs in large stationary sources like electric power plants and refineries (e.g., Acid Rain Program in the US, RECLAIM Program in Southern California). It should not be surprising then, that environmental authorities continue relying on command-and-control instruments (i.e., standards) to regulate emissions from smaller sources because compliance with such instruments only requires the authority to ensure that the regulated source has installed the required abatement technology or that its emissions per unit of output are equal or lower than a certain emissions rate standard.

It appears then that pollution markets are not suitable for effectively air pollution in large cities (e.g., Santiago-Chile, Mexico City, Sao Paulo-Brazil) where emissions come from many small (stationary and mobile) sources rather than a few large stationary sources. Rather than disregard pollution markets as a policy tool, I think the challenge faced by policy makers in cities suffering similar air quality problems is when and how to implement these markets using monitoring procedures that are similar to those under CAC regulation. While the literature provides little guidance on how to approach this challenge (Lewis, 1996), it is interesting to observe that despite its incomplete information on each source's actual emissions, Santiago-Chile's environmental agency has already implemented a market to control total suspended particulate (TSP) emissions from a group of about 600 stationary sources (Montero *et al.*, 2002). Based on estimates from annual inspection for technology parameters such as source's size and fuel type, Santiago's environmental regulator approximates each source's actual emissions by the maximum amount of emissions that the source could potentially emit in a given year.

Motivated by Santiago's TSP permits program, in Montero (2006) I develop a model to precisely study the implications that imperfect monitoring can have on the design of a permits program and on when and whether a permits program ought to be preferred to a conventional standards regulation. The model extends the basic model in different directions. I add a competitive market for an homogeneous good supplied by the same polluting firms. Thus, each firm produces output y and emissions e of a (uniform flow) pollutant. When the firm does not utilize any pollution abatement device $e=y$. A firm can abate pollution at a positive cost by installing technology z, which reduces emissions from $e=y$ to $e=(1-z)y$. Hence, the firm's emission rate is $e/y=1-z$. The regulator only observes z, he neither observes y nor e. In addition, each firm has private information on abatement and production costs. These costs vary across firms according to some joint distribution function that it is assumed to be known by the regulator.

The model developed in Montero (2006) allows us to understand the conditions under which the permits policy is welfare superior to the standards (i.e., CAC) policy. On the one hand, the permits policy retains the well known cost-effectiveness property of conventional permits programs, namely that permit trading allows heterogeneous firms to reduce their abatement and production costs. On the other hand, the permits policy can sometimes provide firms with incentives to choose combinations of output and abatement technology that may lead to higher aggregate emissions than under standards. Thus, when (abatement and production) cost heterogeneity across firms is large, the permits policy is likely to work better. In contrast, as heterogeneity disappears, the advantage of permits reduces, and standards might work better provided that they lead to lower emissions.

Permits can result in higher emissions in two situations, and hence, can be potentially welfare dominated by standards. The first case is when firms with relatively large output *ex-ante* (i.e., before the regulation) are choosing low abatement (i.e., the case with a negative correlation between production and abatement costs). When this is so, permits induce primarily low-output firms to install abatement technologies, while standards force all firms to invest in abatement. The second case is when firms doing more abatement find it optimal to reduce output *ex-post*. In this case, the permits policy has the disadvantage of reducing the output of firms installing more abatement technologies relative to the output of those doing less abatement. This problem is less significant under standards because all firms are required to install similar technologies.

One can notice that the instrument choice problem studied in Montero (2006) is similar in spirit and approach to the instrument choice dilemma considered by Weitzman (1974), but there are important differences. Weitzman (1974) compares the relative advantage of a price instrument (taxes) over a quantity instrument when the regulator has imperfect information about the aggregate abatement cost curve (and possibly about the damage curve as well). Thus, cost heterogeneity across firms plays no role in Weitzman's analysis. Instead, in Montero (2006) I compare the performance of two different quantity instruments and focus on the effect of cost heterogeneity on instrument performance. Since I ignore aggregate uncertainty, the permits policy here is no different than a tax set equal to the permits price and applied over the same proxy emissions.

In deciding whether to use permits or standards, the regulator is therefore likely to face a trade-off between cost-effectiveness and possible higher emissions, so it becomes relevant to discuss the advantages of implementing an optimal hybrid policy in which permits are combined with an optimally chosen standard. While the introduction of the hybrid policy provides a net welfare increase for several combinations of parameters, it is found there are many cases in which the hybrid policy adds nothing over the permits policy (i.e., the hybrid policy converges to the permits-alone policy). Conversely, there are very few cases in which the hybrid policy converges to the standards-alone policy.

Voluntary participation

For either practical or political reasons, phase-in or less than fully comprehensive tradeable permit programs with voluntary opt-in possibilities are attracting considerable attention among policy makers. The Acid Rain Program provides a good example. Under the Substitution provision of this program, electric utility units not originally affected by the program could voluntarily become subject to all compliance requirements of affected units and receive SO_2 tradeable permits approximately equal to their 1988 emissions level (7 years before compliance). Another salient example is provided by current emissions trading proposals in dealing with global warming that call for early carbon dioxide restrictions on OECD countries with (voluntary) substitution possibilities with the rest of the world. Yet another example is provided by trading proposals in dealing with air pollution in Santiago that would allow voluntary participation of non-affected sources (e.g., expansion or creation of parks to sequester PM10).

Although the Substitution provision was primarily designed to allow those non-affected electric units with low abatement cost to (voluntarily) opt-in, Montero (1999) explains that a large number of non-affected units opted in because their unrestricted or counterfactual emissions (i.e., emissions that would have been observed in the absence of regulation) were below their permit allocations. In other words, they had received permits. While shifting reduction from high-cost affected units to low-cost non-affected units reduces aggregate compliance costs, excess permits may lead to social losses from higher emissions than had the voluntary provision not been implemented.

As with any other regulatory practice, the optimal design of a phase-in permits program with opt-in possibilities for non-affected firms is subject to an asymmetric information problem in that the regulator has imperfect information on individual unrestricted emissions and control costs. In world of perfect information (as in the basic model), a regulator would issue permits to opt-in firms equal to their counterfactual emissions. In practice, however, the regulator cannot anticipate the level of counterfactual emissions. Yet, he must establish a permit allocation rule in advance that cannot be changed easily even if new information would suggest so.[7]

As explained by Montero (2000), in deciding how to set the permits allocation rules for affected and opt-in firms, the regulator faces the classical trade-off in regulatory economics (Laffont and Tirole, 1993) between production efficiency (minimization of aggregate control costs) and information rent extraction (reduction of excess permits). In fact, a too restrictive allocation rule for opt-in sources may be effective in controlling the issuance of too many excess permits but at the same time may prove ineffective in attracting low-cost sources. A more generous allocation rule, on the other hand, may be effective in attracting most low-cost possibilities but ineffective in preventing the issuance of excess permits to opt-in sources (with both high and low costs).

To study this regulatory problem, in Montero (2000) I extend the basic model in different directions. First, I consider two groups of firms: affected and non-affected firms. Second, I let the firm's unrestricted emissions or counterfactual emissions be u instead of fixed at 1, which are expected to be equal to historic emissions, that is $E[u]=1$. The actual value of u, however, is the firm's private information which differs across firms according to some density function defined over the interval $[u_l, u_h]$. Third, since abatement cost may differ, on average, across the two groups of firms,[8] I let $c \in [c_l, c_h]$ for affected firms and $c \in [c_l, c_n]$ for non-affected firms, where c_n may be equal to, higher or lower than c_h. The regulator's problem is that of finding permit allocations for affected and opt-in firms that maximizes social welfare subject to imperfect information, cost and benefit uncertainty and design constraints (for example, the definition of the group of firms is assumed beyond the control of the regulator).

One of the results of Montero (2000) is that if the regulator has two instruments – the permit allocation to originally affected firms and to opt-in firms (those non-affected firms that have decided to opt-in) – in the absence of income effects and distributional concerns, the regulator can achieve the first-best outcome. To do so, the regulator sets the permit allocation of opt-in firms high enough (i.e., u_h) such that all non-affected firms opt-in. The total excess permits that are expected to be allocated to opt-in sources (i.e., u_h-1) are deducted from the allocations to affected sources. If the regulator, however, cannot make 'permit transfers' from affected to opt-in sources, so that he has only one instrument – the permit allocation to opt-in firms – he achieves a second-best outcome in which the opt-in allocation is lower than the first-best allocation to the point where the gains from information rent extraction are just offset by the productive efficiency losses of leaving some low-cost non-affected firms outside the program.[9]

Market power

The question of whether and how a firm or a group of firms acting as a cartel can manipulate a pollution permits market has always preoccupied policy makers. The classic work in this area is Hahn (1984) who demonstrates for a static context that a large polluting firm fails to exercise market power when its permits allocation is exactly equal to its emissions in perfect competition. If the permits allocation is above (below) its competitive level of emissions, then, the large firm would find it profitable to restrict its supply (demand) of permits in order to move prices above (below) competitive levels.

In Liski and Montero (2005 and 2006) we revisit this market-power question but in a dynamic context, that is, we allow firms to save today's permits for future use.[10] This banking provision is particularly relevant for permits programs that include a transition phase before reaching the long-run emissions goal. The SO_2 allowance market is a salient example with more than 30 percent of allowances allocated for phase 1 (from 1995 thru 1999) being saved for use in phase 2 (Ellerman and Montero, 2005). The eventual carbon permits market that will develop under the Kyoto Protocol is also likely to make substantial use of the banking provision in anticipation of tighter caps in the future.

Because the evolution of permits market with banking involves the gradual consumption of a stock of permits until the long-term emissions goal is reached, it is natural to borrow from the literature on market power in nonrenewable resources, which started with Salant (1976), to get insights about the properties of the equilibrium path for a permits market dominated by a large agent. Unfortunately Salant (1976) and Hahn (1984) provide somehow conflicting insights. From Salant (1976) we would conjecture that market manipulation would occur independently of the initial permits allocations in that the large firm would always deplete its stock of permits at a rate strictly lower than that under perfect competition. On the other hand, from Hahn (1984) we would conjecture that the type of market manipulation in a dynamic permits market is dependent on the initial allocations in that the rate at which the large firm would (strategically) deplete its stock of permits can be either higher or lower than the perfectly competitive rate.

The properties of our subgame-perfect equilibrium of Liski and Montero (2005) solve the conflict between the two conjectures. Our equilibrium outcome is qualitatively consistent with Salant (1976) if and only if the fraction of the initial stock allocated to the large firm is above some (strictly positive) critical level; otherwise, firms follow the perfectly competitive outcome (i.e., prices rising at the rate of interest up to the exhaustion of the entire permits stock). As in Hahn (1984), the critical level is exactly equal to the fraction of the stock that the large firm would have needed to cover its emissions along the competitive path. But unlike Hahn (1984), the large firm cannot manipulate prices below competitive levels if its stock allocation is below the critical level. We believe that this explicit link between the stock allocation and market power has practical value, since evaluating the critical share of the stock for each participant of the trading program is relatively easy.

The reason our equilibrium can exhibit a sharp departure from the predictions of Salant (1976) and Hahn (1984) is because the large firm must simultaneously solve two opposing objectives (revenue maximization and compliance cost minimization) and at the same time face a fringe of members with rational expectations.[11] At each point in time the large firm decides how many permits to bring to the spot permits market and how much stock to leave for the next periods (recall the large firm also consumes permits for its own compliance). Fringe members clear the spot market and decide their remaining stocks based on what they (correctly) believe the market development will look like.

When the initial stock allocated to the large firm is large enough (i.e., above the critical level), the equilibrium path is governed by two equilibrium conditions: the large firm's marginal revenue and marginal cost rise at the rate of interest. As the large firm's initial stock decreases, these two conditions start conflicting with each other and they cannot longer hold when the initial stock is below the critical level (it only holds that marginal cost grow at the rate of interest). Furthermore, when the stock is below the critical level, the equilibrium cannot depart from competitive pricing; otherwise, fringe members will be holding no stocks, which would contradict their rational expectations.

We are in the process of calibrating this theoretical framework to simulate the development of a carbon market in which a large player, say Russia, plays against a fringe containing all the other countries affected by the Kyoto Protocol.

CONCLUSIONS

I have extended the basic model of pollution control under perfect information to accommodate topics that seem relevant for the practical design and implementation of pollution markets. Either for space constraints or limited literature, several topics have been left out. Let me mention a few. The first is whether the initial allocation of permits makes much of a difference on the performance of the market and on overall welfare. A free allocation of permits (contingent on source operation) may induce too much entry from a long-run perspective which does not happen when they are auctioned off (Spulber, 1985). In the presence of pre-existing tax distortion (e.g., labor and capital taxes), a free allocation of permits may also be welfare inferior to auctioning them off (Goulder *et al.*, 1997).

A second important topic that has attracted considerable attention in the global warming discussion is the effect of regulation on technological change. Market-based instruments such as permits are taxes are generally believed to provide firms with more incentives to innovate and adopt newer technologies than traditional standards regulation (e.g., Jung *et al.*, 1996). However, such a view has been somewhat challenged recently (e.g., Montero, 2002b). More empirical analysis is needed here.

We also need to start thinking about how to use the 'new empirical IO' methods for structuring empirical tests of market power for permits markets in which firms' costs is private information.[12] It is not obvious to me how to go about this because the demand curve in a permits market is not exogenously given as in a conventional market. The same firms may be sellers of permits today and buyers of permits tomorrow as market conditions change. Furthermore, being a buyer or seller of permits may also depend on whether the market is competitive or not, which is what we need to estimate in the first place. Perhaps information on individual emissions and allocations (which are readily available for compliance reasons) will help to simplify the endeavor.

Other topics not covered in this review article include the design of permits markets for non-uniformly mixed pollutants (O'Ryan, 1996), the welfare implications of allowing firms to trade permits intertemporally (Ellerman and Montero, 2005), the welfare comparison between permits and standards when the regulator cannot set emission targets optimally (Oates *et al.*, 1989), and the design of permit markets in a few players context and where emissions (effort) are imperfectly monitored at the individual level but not at the aggregate level. Further research on this latter topic is particularly relevant if we want to introduce permit markets for water pollution control. The literature on moral hazards in teams pioneered by Holmstrom (1982) should be the starting point.

NOTES

* The article was completed while the author was visiting Harvard's Kennedy School of Government under a Repsol–YPF Fellowship.

1. In a later paper Spulber (1988) argues that the first-best may not be feasible under budget constraints.

2. In a subsidy scheme, the government pays firms for reductions.

3. Note that if there are only two possible realizations of cost (high and low), the introduction of a tax and subsidy implement the first–best.

4. Alternatively, we can simply say that the regulator lacks sufficient resources to induce full compliance.

5. To make sure that a non-compliant firm found submitting a false report is in compliance during the next period (but not necessarily the period after), we can assume that the regulator always inspects the firm during that next period, and in the case the firm is found to be out of compliance again, the regulator raises the penalty to something much more severe.

6. Noncompliance and truth-telling could be a feasible strategy if firms reporting noncompliance were subject to a fine lower than F. See Kaplow and Shavell (1994) and Livernois and McKenna (1999) for details.

7. Instead of using an allocation rule, one can work on a case-by-case basis, which most certainly would make the opt-in process more costly for both the regulator and for firms.

8. In the global warming case, it is likely that carbon abatement costs of sources affected by the Kyoto Protocol are, on average, significantly higher than the costs of non-affected sources (i.e., less developed countries).

9. Montero (2000) also found that the second-best result is sensitive to uncertainty in benefits and aggregate control cost. In fact, if benefit and cost uncertainties are correlated negatively or not at all, the regulator benefits from setting the opt-in rule slightly above the 'certain' second-best allocation.

10. Hagem and Westskog (1998) also attack this question but their two-period model is unable, by construction, to capture some of the market developments we characterize.

11. Salant (1976) also considers a fringe with rational expectations but the large firm's decision at any period reduces to one variable: the amount of oil to bring to the market in that period. In Hahn (1984), the large firm's decision also reduces to one variable: the amount of permits to bring to the market.

12. See Bresnahan (1989) for more on the new empircal IO literature.

REFERENCES

Baumol, W.J. and W.E. Oates (1988), *The Theory of Environmental Policy*, Cambridge University Press, Cambridge, UK.

Becker, G.S. (1968), 'Crime and punishment: An economic approach', *Journal of Political Economy*, **76**, 169-217.

Bresnahan, T.F. (1989), Empirical studies of industries with market power. In R. Schmalensee and R.D. Willig, eds, *The Handbook of Industrial Organization*, New York: North-Holland.

Dasgupta, P., P. Hammond and E. Maskin (1980), 'On imperfect information and optimal pollution control', *Review of Economic Studies*, **47**, 857-860.

Ellerman, A.D., P. Joskow, R. Schamlensee, J.P. Montero and E.M. Bailey (2000), *Markets for Clean Air: The US Acid Rain Program*, Cambridge University Press, New York.

Ellerman, A.D. and J.P. Montero (2005), 'The Efficiency and robustness of allowance banking in the US acid rain program', working paper 2005-05, MIT-CEEPR.

Foster, V. and R.W. Hahn (1995), 'Designing more efficient markets: Lessons from Los Angeles smog control', *Journal of Law and Economics*, **38**, 19-48.

Goulder, L.H, I.W. Parry and D. Burtraw (1997), 'Revenue-raising versus other approaches to environmental protection: The critical significance of preexisting tax distortions', *RAND Journal of Economics*, **28**, 708-731.

Hagem, C. and H. Westskog (1998), 'The design of a dynamic tradable quota system under market imperfections', *Journal of Environmental Economics and Management*, **36**, 89-107.

Hahn, R.W. (1989), 'Economic prescriptions for environmental problems: How the patient followed the doctor's orders', *Journal of Economic Perspectives*, **3**, 95-114.

Hahn, R.W. (1984), 'Market power and transferable property rights', *Quarterly Journal of Economics*, **99**, 753-765.

Holmstrom, B. (1982), 'Moral hazard in teams', *Bell Journal of Economics*, **13**, 324-340.

Jung, C., K. Krutilla and R. Boyd (1996), 'Incentives for advanced pollution abatement technology at the industry level: An evaluation of policy alternatives', *Journal of Environmental Economics and Management*, **30**, 95-111.

Kaplow, L. and S. Shavell (1994), 'Optimal law enforcement with self reporting of behavior', *Journal of Political Economy*, **102**, 583-606.

Kwerel, E. (1977), 'To tell the truth: Imperfect information and optimal pollution control', *Review of Economic Studies*, **44**, 595-601.

Laffont, J.J. and J. Tirole (1993), *A Theory of Incentives in Procurement and Regulation*, Cambridge, Massachusetts: MIT Press.

Lewis, T. (1996), 'Protecting the environment when costs and benefits are privately known', *RAND Journal of Economics*, **27**, 819-847.

Liski, M. and J.P. Montero (2006), 'On pollution permit banking and market power', *Journal of Regulatory Economics*, **29**, 283-302.

Liski, M. and J.P. Montero (2005), 'Market power in a storable-good market: Theory and applications to carbon and sulfur trading, working paper', CEEPR, MIT.

Livernois, J. and C.J. McKenna (1999), 'Truth or consequences: Enforcing pollution standards with self-reporting', *Journal of Public Economics*, **71**, 415-440.

Montero, J.P. (2006), 'Pollution markets with imperfectly observed emissions', *RAND Journal of Economics*, **36**, 645-660.

Montero, J.P. (2002a), 'Prices vs. quantities with incomplete enforcement', *Journal of Public Economics*, **85**, 435-454.

Montero, J.P. (2002b), 'Permits, standards, and technology innovation', *Journal of Environmental Economics and Management*, **44**, 23-40.

Montero, J.P. (2001), 'Multipollutant markets', *RAND Journal of Economics*, **32**, 762-774.

Montero, J.P. (2000), 'Optimal design of a phase-in emissions trading program', *Journal of Public Economics*, **75**, 273-291.

Montero, J.P. (1999), 'Voluntary compliance with market-based environmental policy: Evidence from the US Acid Rain Program', *Journal of Political Economy*, **107**, 998-1033.

Montero, J.P., J.M. Sánchez and R. Katz (2002), 'A market-based environmental policy experiment in Chile', *Journal of Law and Economics*, **45**, 267-287.

Montgomery, W.D. (1972), 'Markets in licenses and efficient pollution control programs', *Journal of Economic Theory*, **5**, 395-418.

Oates W. E., P.R. Portney and A.M. McGartland (1989), 'The net benefits of incentive based regulation: A case study of environmental standards setting', *American Economic Review*, **79**, 1233-1242.

O'Ryan, R. (1996), 'Cost-effective policies to improve urban air quality in Santiago, Chile', *Journal of Environmental Economics and Management*, **31**, 302-313.

Pizer, W. (2002), 'Combining price and quantity controls to mitigate global climate change', *Journal of Public Economics*, **85**, 409-434.

Roberts, M. and M. Spence (1976), 'Effluent charges and licenses under uncertainty', *Journal of Public Economics*, **5**, 193-208.

Salant, S.W. (1976), 'Exhaustible resources and industrial structure: A Nash-Cournot approach to the world oil market', *Journal of Political Economy*, **84**, 1079-1093.

Schmalensee, R., P. Joskow, A.D. Ellerman, J.P. Montero and E.M. Bailey (1998), 'An interim evaluation of sulfur dioxide emissions trading', *Journal of Economic Perspectives*, **12**, 53-68.

Spulber, D. (1988), 'Optimal environmental regulation under asymmetric information', *Journal of Public Economics*, **35**, 163-181.

Spulber, D. (1985), 'Effluent regulation and long-run optimality', *Journal of Environmental Economics and Management*, **12**, 103-116.

Stavins, R. (2003) 'Experience with market-based environmental policy instruments', in K.G. Maler and J. Vincent (eds), *Handbook of Environmental Economics*, Amsterdam: Elsevier Science.

Tietenberg, T. (2005), *Emissions Trading: Principles and Practice*, Washington, D.C.: Resources for the Future.

Tietenberg, T. (1985), *Emissions Trading: An Exercise in Reforming Pollution Policy*, Washington, D.C.: Resources for the Future.

Weitzman, M. (1974), 'Prices vs. quantities', *Review of Economic Studies*, **41**, 477-491.

PART II

The EU Emission Trading System

4. European greenhouse gas emissions trading: A system in transition

John Reilly and Sergey Paltsev

INTRODUCTION

The United Nations Framework Convention on Climate Change (UNFCCC) was ratified by 154 countries in 1992 with the ultimate objective to achieve stabilization of greenhouse gas concentrations in the atmosphere at a level which prevents dangerous interference with the climate system. In 1997 the Kyoto Protocol to the UNFCCC was adopted, where a set of industrialized countries agreed to limit their greenhouse gas emissions. The Protocol entered into force in 2005 imposing emission limits for 2008 to 2012. In negotiations leading up to the Protocol, the US was the leading proponent of international emissions trading, with the European Union initially reluctantly agreeing to this inclusion. The US withdrawal from the Protocol has greatly changed the nature of the agreement. One interesting turn is that the European Union has now fully embraced emissions trading, establishing in 2003 (EC, 2003) the Emission Trading Scheme (ETS). It will run for the three year period of 2005 to 2007 with the intent of helping to prepare its member states to achieve compliance with their international commitments in 2008 to 2012 under the Kyoto Protocol. This is the first serious effort anywhere in the world to establish a cap and trade system for greenhouse gas emissions and the performance of the system is being widely watched.

The ETS was designed as a test system, and in so doing the goal from the start was to establish a relatively mild reduction requirement so that basic operations such as establishing registries and becoming familiar with the trading instrument could occur while the financial stakes were relatively low. Establishing the National Allocation Plans (NAPs) for countries has been a difficult process of negotiations between member states and the European Commission which had final approval. Just how binding the allocations would be has only gradually come into focus, but based on the estimates of the member states themselves the reductions for major countries approved early in the process was in the order of 1 percent below projected reference emissions. Countries, whose NAPs were submitted and approved later, eventually fell in line with early submitters, sometimes under pressure from the EC to reduce allocations where it was argued that initial NAPs would convey unfair competitive advantage to firms in some of the member states. Thus most analysts concluded the ETS caps would be hardly binding and the carbon prices would be very low. One attempt to measure expected carbon prices reported a median expectation of 5.50€/tCO$_2$ with a low and high range of 2.50 and 10.00€/tCO$_2$ (Pew Center, 2005, reporting results of an ongoing survey of expected prices conducted by Point Carbon). This was the expectation reported as of December 2003 before the system went into effect. Median expectations as of April 2005 reported by the same survey, when the market price was in the range of 15 to 18€/tCO$_2$, remained at 7.00€/tCO$_2$ although the high end expectation was 45€/tCO$_2$. Thus, observers, at least those represented in this survey, remained somewhat sceptical that the relatively high price would be supported over the longer term. One early study (See, 2005) used Monte Carlo analysis to estimated probability density function for the permit price for the ETS. Under several variants of the Monte Carlo analysis, he found the median carbon price to be under 0.5€/tCO$_2$ with a maximum price over all variants less than 7€/tCO$_2$. This study was concluded before recent downward adjustments in some countries caps.

The actual trading prices under the ETS thus have been a surprise for analysts, settling in around 20 to 25€/tCO$_2$ (~70 to 90 €/tC) by mid- to late 2005 having approached 30€/tCO$_2$ (110€/tC) in July. To put these prices in context, it is useful to contrast these with projections of the carbon price required to meet the Kyoto Protocol in its early versions before the US withdrew. At this stage, the Protocol initially envisioned that Parties would by 2008 to 2012 reduce emissions to 5 percent below 1990 levels. With continued economic growth this was widely projected to be a reduction on the order of 20 percent from reference emissions (i.e., what emissions would have been by that time in the absence of mitigation efforts). A comparison of several key models showed that 7 of 11 models estimated the carbon price needed to meet the approximately 20 percent Kyoto cut to be in the range of 20-35€/tCO$_2$[1] (Weyant and Hill, 1999), which is about the current trading price range for the ETS. One of the studies in that comparison estimated a lower price, and three estimated a higher carbon price but nothing that would suggest that a one percent cut might lead to a price of 20€/tCO$_2$ or more.

The ETS is still evolving. At the time of writing we are only partway through the first year of the three-year program, rules are still being defined, registries are still being established, market participants have little experience with the permit trading, the volume of trade is small, and expectations about the future of emissions trading in Europe beyond the ETS may be driving current prices. Further information about current ETS structure and issues related to carbon allocations can be found in Reilly and Paltsev (2005). The EC maintains the web site (http://europa.eu.int/comm/environment/climat/emission_plans.htm) with links to countries' NAPs. More information on the initial assessment of NAPs can also be found in Betz *et al.* (2004) and Zetterberg *et al.* (2004).

This chapter is an early attempt to contrast projections of carbon prices in the ETS period with actual prices to date, and speculate on what could explain the huge gap. The chapter is organized as follows. Section 2 describes the version of the EPPA model used here. Section 3 first reports the results of the central EPPA projections, and then we speculate on reasons for the gap between these results and market prices, supplementing this speculation with additional model analysis where possible. Section 4 offers some conclusions and final thoughts.

EPPA–EURO MODEL

The ETS establishes a framework for trading in carbon dioxide (CO_2) emissions across the original EU-15 nations and the 10 accession countries (Table 4.1). The ETS runs from 2005 to 2007 and covers large emitters in the power and heat generation and in selected energy-intensive industrial sectors: combustion plants, oil refineries, coke ovens, iron and steel plants and factories making cement, glass, lime, bricks, ceramics, pulp and paper. To analyze the ETS, we apply the MIT Emissions Prediction and Policy Analysis (EPPA) model (Babiker *et al.*, 2001, Paltsev, *et al.*, 2005). EPPA is a recursive-dynamic multi-regional general equilibrium model of the world economy. The EPPA model is part of a larger Integrated Global Simulation Model (IGSM) that predicts the climate and ecosystem impacts of greenhouse gas emissions (Sokolov *et al.*, 2005), but for this study is run in stand-alone mode, without the full IGSM.

The EPPA model is built on the GTAP data set, which accommodates a consistent representation of energy markets in physical units as well as detailed accounts of regional production and bilateral trade flows. Besides the GTAP data set, EPPA uses additional data for greenhouse gas (CO_2, CH_4, N_2O, HFCs, PFCs, and SF_6) and urban gas (SO_2, NO_x, CO, NH_3, VOC, black

carbon, and organic carbon) emissions. For use in the main version of the
EPPA model the GTAP dataset is aggregated into the 16 regions and 10 sectors
(Paltsev *et al.*, 2005). In order to represent the ETS in the EPPA model, we
introduce additional regional disaggregation, where Europe (EUR) and
Eastern Europe (EET) are disaggregated into 12 EU regions (Table 4.2) and a
block of non-EU European countries. We call this version of the model as
EPPA-EURO.

*Table 4.1   National allocation plan CO_2 caps, 2003 CO_2 emissions, and the
Kyoto Protocol targets*

	2005–2007 allowance for ETS sectors, mmt	Average Annual allowance for ETS sectors, mmt	2003 national CO_2 emissions, mmt	Share of ETS allocated emissions in 2003 total national emissions	Kyoto target, percent relative to baseyear
Austria	99.01	33.0	76.2	0.43	-13
Belgium	188.8	62.9	126.3	0.50	-7.5
Cyprus	16.98	5.7	7.2	0.79	No target
Czech Republic	292.8	97.6	127.1	0.77	-8
Denmark	100.5	33.5	59.3	0.56	-21
Estonia	56.85	19.0	19.1	0.99	-8
Finland	136.5	45.5	73.2	0.62	0
France	469.53	156.5	408.2	0.38	0
Germany	1497	499.0	865.4	0.58	-21
Greece	223.2	74.4	110	0.68	+25
Hungary	93.8	31.3	60.5	0.52	-6
Ireland	67	22.3	44.4	0.50	+13
Italy	697.5	232.5	487.3	0.48	-6.5
Latvia	13.7	4.6	7.4	0.62	-8
Lithuania	36.8	12.3	12.3	1.00	-8
Luxemburg	10.07	3.4	10.7	0.31	-28
Malta	8.83	2.9	2.5	1.18	No target
Netherlands	285.9	95.3	176.9	0.54	-6
Poland	717.3	239.1	321.3	0.74	-6
Portugal	114.5	38.2	64.3	0.59	+27
Slovakia	91.5	30.5	43.1	0.71	-8
Slovenia	26.3	8.8	16.1	0.54	-8
Spain	523.7	174.6	331.8	0.53	+15
Sweden	68.7	22.9	56	0.41	+4
UK	736	245.3	557.5	0.44	-12.6
EU-15	5217.9	1739.3	3447.5	0.50	-8
EU-25	6572.8	2190.9	4064.1	0.54	

Source for allowances data: for Poland, Greece, Italy, and Czech Republic – their NAPs and the
EU Commission Decisions available at: http://europa.eu.int/comm/environment/climat/emission_
plans.htm; for the other countries – EC (2005).
Source for emissions data: EEA (2005).

The Kyoto Protocol targets are percentage changes in GHG emissions for 2008–2012 relative
to base year levels. The target is for all six GHG (not just CO_2) and is expressed in terms of CO_2
equivalence. For Finland and France, the base year is 1990 for emissions of all GHGs. For the
other EU-15 countries, the base year is a combination of 1990 emissions of CO_2, CH_4, and N_2O,
and 1995 emissions of HFC, PFC and SF_6.

Table 4.2   *EU regional aggregation in the EPPA–EURO model and the ETS*
            allocation: ratio of allocated emissions in electricity (ELEC) and
            energy-intensive industries (EINT) to projected emissions in 2005

EPPA–EURO Region	Countries	ELEC	EINT
FIN	Finland	1	1
FRA	France	1	0.99
DEU	Germany	0.99	0.99
GBR	UK	0.99	0.99
ITA	Italy	0.99	0.99
NLD	Netherlands	1	1
ESP	Spain	0.936	0.946
SWE	Sweden	0.86	0.86
REU	Austria, Belgium, Cyprus, Denmark, Estonia, Greece, Ireland, Latvia, Lithuania, Luxemburg, Malta, Portugal	0.98	0.98
HUN	Hungary	1.05	1.05
POL	Poland	1.01	1.01
XCE	Czech Republic, Slovakia, Slovenia	1.01	1.01

Source: National Allocation Plans and Betz et al. (2004).

Note: final ETS allocations across sectors have tighter constraints on ELEC sector and relaxed
       constraints on EINT sector than shown here. It does not affect the analysis because sectors
       can trade freely.

The base year of the EPPA model is 1997. From 2000 onward it is solved
recursively at five year intervals. Because of the focus on climate policy, the
model further disaggregates the GTAP data for energy supply technologies
and includes a number of backstop energy supply technologies that were not
in widespread use in 1997 but could take market share in the future under
changed energy price or climate policy conditions. The EPPA model
production and consumption sectors are represented by nested Constant
Elasticity of Substitution (CES) production functions (or the Cobb-Douglas
and Leontief special cases of the CES). The model is written in GAMS-
MPSGE. It has been used in a wide variety of policy applications (see Paltsev
et al. (2005) for a list of EPPA applications).

We apply NAP allocation caps in EPPA as if they are national caps where
only the two sectors are participating in emissions trading. Our BAU for ETS
sectors are presented in Table 4.2, as a ratio of allocated to projected emissions
for 2005. In economic theory, what matters in terms of trading and economic
efficiency is the market clearing permit price. That is, even if a firm were

given enough permits to cover its emissions (and thus could comply without abating), economic theory would argue that the firm would operate on the opportunity cost of carbon emissions – if it could abate at or below the market price it could sell excess allowances at the market price. Further, prices of goods should reflect the marginal cost of production, which would include the marginal cost of abatement. A large allocation of permits to a firm is a lump sum distribution, which according to theory, would enrich the firm's shareholders but would not affect operating decisions or competitiveness. A competing firm that got few allowances would suffer a relative loss, but again this would be a one time loss due to the small lump sum allocation, and it would not further affect operating decisions. CGE models like EPPA follow this neoclassical economic theory closely. Thus, how permits are allocated does not affect which sectors or firms abate or production decisions even if they are given away for free. The cap and trade system is thus modelled as if all permits were purchased from the government and all revenue is distributed in a lump sum manner to the representative consumer. Neoclassical economic theory would show the allocation to affect the distribution of income, depending on the extent to which different consumers own equity of firms allocated portions of the cap or affected by it (directly or indirectly), consume goods whose prices are affected by the cap, and are employed by firms directly or indirectly affected by the cap. Since EPPA has a single consumer who owns all assets and supplies all labor, it does not provide any direct information on the distributional effects. We also cannot estimate the potential distortionary effects of non-lump sum distribution of some of the permits (those that under some countries' NAPs are retained for new entrants).

We note other approximations and caveats: (1) By including sectors as a whole, we are unable to represent the exclusion of small sources, and this represents a potential avenue of leakage and inefficiency to the extent the ETS encourages production to shift to small sources. (2) We model interaction with existing energy taxes, elsewhere showing (Paltsev *et al.*, 2004) that this strongly affects economic impacts of a cap and trade system, although not the carbon price. Notably, fuel taxes are relatively low in the sectors capped under the ETS. However, the issue of interaction of multiple policies is an issue of importance given that there are a variety of other policies directed toward the capped sectors, such as targets for wind power and renewable energy in electricity production. Since our interest here is exclusively on the simulated carbon price, these broader economic effects that would be captured in other measures of cost are less relevant. (3) EPPA solves every 5 years, and we have thus taken the year 2005 as illustrative of the 2005 to 2007 ETS period and 2010 as illustrative the 2008 to 2012 Kyoto period. However, we attempt to correct for not having the mid-year of the ETS period by calculating the cap as a percentage below reference projections for emissions over the period 2005–2007 and applying this to 2005. Thus, we approximate the average reduction required over the period to the extent these sectors are projected to grow even though the simulation is for 2005. (4) Business-As-Usual forecasts are key determinants of the carbon price. As illustrated in Reilly and Paltsev (2005), there can be large year-to-year changes in emissions for countries (both positive and negative). For example, between 2002 and 2003 Finland's emissions from electricity and industry grew by 20 percent, while in Portugal they fell by 14 percent. These big changes reflect availability of hydroelectricity, changes in fossil fuel prices, and other factors that can be highly variable from year-to-year. Such variability is generally not captured in a model such as EPPA, where any one year simulation should be more properly interpreted as a multi-year average result.

RESULTS

In order to evaluate the likely development of carbon price in the ETS, we have considered the scenarios presented in Table 4.3, where we allow for different trading regimes across the EU countries. Scenario 1 illustrates the range of prices in the cases of no carbon trading among countries. This is a useful way to judge the extent to which caps are more or less binding in different countries. Scenario 2 is the closest to the current ETS design and our BAU projections. In Scenario 3 we have eliminated remaining 'hot air' from Eastern European countries. The projected carbon prices are presented in Table 4.4. Scenario 1 shows that most of the original EU-15 member states have caps that result in similar carbon prices of generally at or below 1 €/tCO$_2$ with the exception of Sweden, Spain, and Italy where autarkic carbon prices are just over 15, 6, and 2 €/tCO$_2$ respectively. In contrast, there is 'hot air' in the newly admitted EU countries of Poland, Hungary, and our aggregate of the remaining countries of Eastern Europe, and the autarkic price in these areas is zero. Trading across the EU equalizes the carbon price at 0.58 €/tCO$_2$ (Scenario 2). Eliminating the 'hot air' in the newly admitted countries by setting the cap at reference emissions in these areas increases the price to 0.85 €/tCO$_2$ (Scenario 3).

Table 4.3   ETS scenarios for 2005–2007

Scenario 1	Carbon trading across sectors within countries but not across countries.
Scenario 2	Carbon trading across sectors and countries (allowing ETS allocation above BAU projections).
Scenario 3	Carbon trading across sectors and countries but no countries' sectors get more allowances than reference emissions (no hot air).

Table 4.4   Carbon price in different ETS scenarios

Region	Scenario 1, €/tCO$_2$	Scenario 2, €/tCO$_2$	Scenario 3, €/tCO$_2$
FIN	0.14	0.58	0.85
FRA	0.30	0.58	0.85
DEU	0.85	0.58	0.85
GBR	0.96	0.58	0.85
ITA	1.79	0.58	0.85
NLD	0.16	0.58	0.85
ESP	6.16	0.58	0.85
SWE	15.24	0.58	0.85
REU	2.47	0.58	0.85
HUN	0.00	0.58	0.85
POL	0.00	0.58	0.85
XCE	0.00	0.58	0.85

Note: the ratio of a price per ton of CO$_2$ to a price per ton of carbon is 1:3.667 based on a carbon content of CO$_2$.

These simulated prices are completely at odds with observed ETS market prices that have been in the range of 20 to 25€/tCO$_2$. A number of theories or factors have been advanced to explain the unexpectedly high prices. These include:

1. Increases in energy prices (gas and oil) caused a shift to coal use especially in electric generation, which has higher carbon emissions.
2. Recent experience has emphasized the potential effects of adverse weather conditions (drought and high temperatures) on hydro and even on nuclear supply. Drought reduced hydro capacity and high temperatures have led to concerns that discharged cooling water from nuclear power installations could lead to exceedance of in-stream water temperature limits set to avoid damage to these freshwater ecosytems.
3. Expectations regarding the future evolution of emission trading beyond the 2005 to 2007 period. Banking of allowances to future periods would be one

way that expectations about 2008 to 2012 or beyond could affect current ETS prices. France and Poland allow for a limited banking into the Kyoto period, but it is not clear if such banking will be allowed by the European Commission. Another consideration advanced by some analysts is that companies may believe that baseline allocations in 2008 to 2012 will be benchmarked to actual emissions in the ETS years. This would provide an incentive not to abate now to ensure a larger allocation in future years.

4. The EPPA model (as other CGE models) may represent abatement as too easy. The model does not represent accurately the details of the market design, and it does not include transactions costs.

5. The current market prices for carbon do not reflect supply and demand interactions: confusion, speculation, incomplete registries, bad information, or manipulation of the market may be having an effect, particularly as the market gets started.

We further discuss and investigate these issues, in turn.

High Natural Gas and Oil Prices

Dispatching gas generation capacity while cutting back on the dispatch of coal capacity can reduce CO_2 emissions by more than half because the fuel specific release of CO_2 from gas is only about 60 percent of the release from coal, and gas generation, particularly from combined cycle facilities, can be more than twice as efficient (electricity produced/energy content of fuel) as a base load coal plant. Some analysts have calculated the cost of this option as gas prices have risen, and found that it could explain the high carbon prices if this were the marginal abatement option. We investigate this consideration with some additional EPPA runs.

The Business-As-Usual EPPA projections already had oil and gas prices approximately doubling from the base year 1997 level, with coal prices little changed. As of mid- to late 2005, fuel prices were considerably higher than this base EPPA projection, with crude oil at over \$60 barrel and gas prices around 8€ per million BTUs (natural gas prices are even higher in the US reaching \$14-15 per million BTUs). These are 3 to 4 times or more the 1997 level. In standard EPPA simulations fuel prices are endogenously determined, however, the model includes the capability to exogenously set prices. We have used this facility to exogenously set fuel prices to examine the impact on the simulated carbon price.

Table 4.5, column 1, shows the carbon price results when oil and gas prices are at 2, 3, 6, and 50 times the 1997 level in 2005, imposed under conditions of Scenario 3 (no 'hot air' in the new EU members). The 50 times the 1997 level is an obviously extreme value, intended to demonstrate the sensitivity of the model over a very wide range. The 2 times 1997 level is, as expected, nearly identical to the BAU case where oil and gas prices are endogenous. Higher oil and gas prices lead to higher carbon prices, rising to about 1.6 and 3.9€/tCO$_2$. At the extreme of 50 times 1997 oil and gas prices the estimated carbon price rises to about 16€/tCO$_2$, still less than recent market prices. The EPPA model includes a discrete NGCC technology, and so we could see the gas-coal margin reflected directly in the carbon price, however, EPPA generally represents abatement possibilities as a continuous response determined by substitution elasticities. EPPA simulates reductions in energy use, stemming from increased prices as an important abatement avenue that a simple technology cost comparison typically does not include. Not only do direct users of fuel reduce fuel use, but users of products produced from fuels (e.g., electricity) also have an incentive to use less of the good. However, if electricity prices are regulated, are based on average costs, or otherwise fail to adjust to pass through higher marginal costs associated with carbon prices, this avenue may be overestimated in EPPA. Reducing this adjustment, however, would not come close to explaining the difference between simulated and actual prices.

Table 4.5   Gas and oil price effects on the carbon price

Increase in gas and oil prices in 2005 relative to 1997	Carbon price with oil and gas price increase only, €/tCO$_2$	Carbon price with oil and gas price increase and 20 percent reduction in hydro and nuclear production, €/tCO$_2$
2	0.85	1.19
3	1.55	2.11
6	3.89	4.94
50	14.97	18.96

Restricted Hydro and Nuclear Production

Unusual weather in 2005 led to low production of hydro electricity that was largely unanticipated. To examine this factor, we restrict nuclear and hydropower to 20 percent below our reference projection for these sources. We then simulate these reductions in combination with the various oil and gas price scenarios to see the effect on the carbon price. The simulations show a 26 to 40 percent increase in the carbon price, with the larger percentage (but smaller absolute) increases occurring at lower oil and gas prices (Table 4.5, column 2).

These experiments show that reduced nuclear and hydro capacity, even in combination with higher oil and gas prices, do not allow us to simulate the current levels of market prices for carbon. There are two important considerations that limit the reality of our simulations. One consideration is that the high fuel prices and reduced hydro and nuclear capacity were unanticipated shocks but the EPPA model simulations produce results whereby firms would have had some time to adjust. While EPPA is not a perfect foresight model (that would imply full knowledge of the shocks ahead of time) the values of elasticities of substitution in EPPA reflect medium-run estimates. EPPA vintages capital, restricting substitution substantially, but only a portion of capital is vintaged, again allowing implicitly some retrofitting. This would lead one to conclude that EPPA simulations would underestimate the effect of an unanticipated shock. A second consideration, however, is that by using 2005 as representative of the full 2005 to 2007 period, we implicitly assume that the 2005 conditions (including the higher fuel prices and reduced hydro and nuclear capacity) persist over the entire period. The ability to borrow allowances should provide the capability of firms to even out such effects. This depends, of course, on firms believing that these are unusual conditions that will not persist over the full ETS period. The belief that these will persist or worsen could explain higher carbon prices than we have simulated.

Expectations for Emissions Trading beyond 2007

As already noted, there are at least two ways future periods could affect prices in the current period. If banking of allowances is allowed, then one might expect over-compliance with current limits to create extra allowances for future periods if one would otherwise expect the carbon price to be substantially higher then. Economic theory would suggest that the discounted expected future price should equal the current price. In general, the ETS disallowed banking into the Kyoto period, but two of the member States included limited banking provisions. In principle, if any one agent could bank allowances, that agent, if its banking levels where unlimited, could by itself bring the expected future discounted future price in line with the current price in a market where allowances were fungible. Such an agent could buy allowances throughout the region, accumulating them until the supply was judged to be sufficient to bring the future price in line with the current price. On the other hand, whether the EC will allow banking even to the limited extent provided for in the French and Polish NAPs is unclear, and the Kyoto Protocol that will define the rules for 2008 to 2012 does not specifically allow banking from an earlier trading system.

The second hypothesis is that firms may expect future allocations to be based on actual emissions in 2005 to 2007. The simple arithmetic of this is as follows. Suppose allowances in 2008 to 2012 are distributed to be 20 percent below actual emissions in 2005 to 2007. Further suppose that a firm has 1000 allowances in the current period of the ETS and faces the decision of abating from 1000 (its emissions if it did nothing) to 500, at an average cost of $11€/tCO_2$, and could sell these in the current market at $21€/tCO_2$. This would look like a profit of 5,000€. However, by our assumption that future allowances are based on actual emissions, the firm's allowances in 2008 to 2012 would be 400 (80 percent of 500) if it abates compared with 800 (80 percent of 1000) if it did not abate. In this example, even if the price in 2008 to 2012 were $20€/tCO_2$ the decision to abate would mean that the firm was giving up allowances worth 8,000€ in the future by abating today. Discounting this 8,000 value back to current at 5 percent shows the value to be just under 6,300€ and so the firm would be nearly 1,300€ ahead by forgoing abatement today and the revenue from the allowance sales. Not considered explicitly here is that 2008 to 2012 is a five year period (whereas 2005 to 2007 is three years), and that the effects of lower allocation may linger into periods beyond 2012.

Table 4.6   Scenarios for 2008–2012 and ETS carbon price

Scenario	Description	Carbon Price, $€/tCO_2$
Scenario 4	ETS extended with unchanged quantity targets in the ETS sectors to 2008–2012, and other sectors are capped to meet Kyoto targets.	13.47
Scenario 5	Kyoto target with trade among all sectors and across EU. Emission trade in CO_2 only.	32.32
Scenario 6	Kyoto target with trade among all sectors and across EU, Canada, Japan, and Russia. Emission trade in CO_2 only.	6.28
Scenario 7	Kyoto target with trade among all sectors and across EU, Canada, Japan, and Russia. Emission trade in all GHGs.	0.70

A critical value in both the banking and the allocation-loss calculations is the expected future price of carbon. We thus construct scenarios in EPPA that represent some different ways in which the ETS could evolve in the 2008–2012 Kyoto period. Table 4.6 presents scenarios and corresponding carbon prices for the Kyoto Protocol period. In Scenario 4 we keep the current ETS sectors and their quantity targets unchanged for 2008 to 2012. This would mean the other sectors of the economy have to reduce their emissions proportionally to meet the Kyoto target, which we have enforced through a cap on these sectors without allowing trading with the ETS sectors. They have different carbon prices (not reported here) than the ETS system, and that has some effects on their demand for goods supplied by the ETS sectors. However, if these two parts of the economy are kept separate, then what matters to the ETS sectors is the price in the emissions market in which they are operating. In Scenario 5 we extend the ETS to all sectors and all EU regions, with the Kyoto targets as allocation caps. The European Commission has expressed a desire to extend the ETS to other sectors, and this is an extreme assumption where it is extended to the entire economy. This scenario does not allow any credits (JI, CDM, trading) from outside of the EU. Scenario 6 expands emissions trading to include the EU, Russia, Canada, and Japan, assuming these other Kyoto Parties will set up national trading systems covering all sectors of their economies. In Scenario 7 such trade is extended to include all greenhouse gases.

If banking on the expectation of higher future prices were an explanation for the high current price, then to support a price of $25€/tCO_2$ we would expect to see the five year undiscounted price higher by about 28 to 47 percent (for a 5 and 8 percent, respectively, discount rate). The future carbon price would thus need to be 32 to $37€/tCO_2$. Scenario 5 results in a price of just over $32€/tCO_2$. Thus anticipation of banking could support the current price if the assumption is that trading will be extended to the other sectors, without any CDM, JI, or trading credits from abroad, and assuming the ETS excludes abatement of non-CO_2 GHGs. This is among the most extreme cases we could construct. Further, since banking is by no means a sure thing one might expect firms to not fully equate the current price to the discounted future expected price because of the risk that a large cache of banked allowances might turn out to be of no value if the EC sticks to its prohibition on banking.

The allocation-loss explanation is possibly more compelling, but it is harder to estimate the full effect. As demonstrated with the simple example, loss of allocation can lead to less abatement even if the future carbon price is no higher than today. Caution is needed in applying this example arithmetic broadly however. The ETS sectors must meet the 2005 to 2007 target assuming the EC strictly enforces the cap, and so to the extent one firm plays a game of not-abating hoping to garner a larger allocation in the next period, other firms will need to abate more. This behaviour would still cause a run-up in the current price, but by how much depends more on the differential abatement opportunities among firms and their other interests in acting strategically.

EPPA Parameterization Underestimates Abatement Cost

If the required abatement is really only on the order of 1 percent below the reference then choice of parameters that affect abatement costs within EPPA would be insufficient to generate carbon prices like those observed in the current ETS market. More compelling than simulating EPPA with changed parameters is the comparison of different model results for a 20 percent reduction that we reported earlier. We thus have not constructed new cases to illustrate this here. See (2005) using the EPPA-EURO model and conducting an uncertainty analysis considered a Monte Carlo case where elasticities of substitution between energy and non-energy inputs were subject to uncertainty and compared results with a case where they were not. He also considered a case where the proportion of capital vintaged was varied with a case where it was not. The effect of varying the elasticity substitution changed the median price by about 5 percent and the maximum price by about 10 percent. The marginal effect of vintaging was smaller on the maximum price, about 7 percent, and larger on the median price, about 15 percent. Neither of these results suggest that changing these parameters could easily explain an increase of an order of magnitude times three, which is what is required to get from 0.85 to 25€/tCO$_2$. If we calculate the arc elasticity that would be needed to have the price rise from 0.85 to 25 €/tCO$_2$, (% Δ Q/% Δ P) for a 1 percent quantity change we get $\{1/[(0.85\text{-}25)/0.85)]\}= 0.035$. Even most short run (one-year) elasticities of substitution are on the order of 0.4 or higher. And the nature of the ETS, with banking and borrowing among ETS years, allows adjustment over three years. A completely different model structure, where there was no flexibility whatsoever at low prices, but one technological option that would kick in once the carbon price reached the level that made it competitive would be more likely to yield a high price even though the required abatement was a trivial percentage of emissions. An example would be if the only near term abatement was natural gas electricity generation substituting for coal. With high gas prices, the trigger point to make this economic could well be on the order of 25€/tCO$_2$. To get this from EPPA, we would need to make the entire economy fixed coefficient, with the only abatement being the technological option of NGCC, an extremely different view of economic response to higher prices than is modelled in EPPA.

Another consideration is that models such as EPPA do not include transactions costs. In this regard, there are many costs to setting up registries and developing inventories within firms, but it is not obvious that these costs would be fully reflected in market prices for permits – both buyers and sellers must bear costs of creating and maintaining inventories and so there is no reason to think that the price would settle at a level where sellers would be compensated for the costs, while buyers must pay. To be sure, this is real additional cost that would be reflected in firms' bottom lines and in prices of goods in the economy but not necessarily in the carbon price. Pure transactions costs, e.g., traders' margins, seem unlikely to result in permit prices that are many multiples of the basic price if the market becomes relatively liquid.

As noted earlier, there are elements of the market design that we have not captured, such as reserved allocations for new entrants, non-functional registries in East European countries, and the fact that if facilities closed down they are required to surrender their allowances. Provisions of the NAPs are still a subject to challenge and this may be affecting market participants' expectations.

Prices Do Not Reflect Market Fundamentals

There is not much more than can be said in this regard, and we hesitate to argue that our model is smarter than the market. We repeat a quote from one trader: 'I am beginning to think there is no real supply-demand indication in this market. It doesn't react to fundamentals'. (Point Carbon, 2005). Experience with emissions trading markets for SO_2 and NO_x shows high volatility, particularly in the early stages. Thus with the ETS in its early stages it is hard to judge whether the short series of prices are representative of what one will observe over the full three years of trading.

CONCLUSIONS

The creation of a carbon market in the European Union is a watershed event in climate policy. How it performs (or as importantly, perceptions about its performance) may well determine whether there is rapid progress toward establishing an international market in permits that could eventually cover much of the world, or the world sours on permit trading and pursues other policy approaches. The EU is an interesting test bed: it is an international market in that the individual EU member states retained some control over National Allocation Plans, but with considerable enforcement power within the European Commission there is a central authority with more power to bring consistency across these plans than would be the case if trying to establish emission trading among the EU, the US, and Japan, or with Russia, China, and India.

Economic theory strongly concludes that creating a cap and trade system for controlling pollutants assures that abatement is achieved in a least cost manner. Experience in the US with such trading systems for other pollutants has been widely seen as highly successful (Ellerman *et al.*, 2000). While there are differences for CO_2 versus other pollutants that may affect how one would like to manage an emissions market, in the main nearly all economists would have a fair amount of faith that decentralized decisions guided by a market price set as an interaction of supply and demand for permits is preferable to command and control systems for pollution control. Economists might argue about other issues related to such a system such as equity, its merits compared with an environmental tax, revenue recycling, interaction with other policies, enforcement in an international regime, and the correct level of a cap. But in terms of cost-effectiveness of such an instrument as exhibited by the marginal market abatement cost, most economists would require strong proof before accepting that a cap and trade system was less effective than some other means of control. That same faith in market instruments may not necessarily hold for non-economists, and so proving that emissions trading can work may not surprise economists but may be essential to garner further support for such mechanisms.

In convincing non-economists of the value of market instruments, perception may count as much as reality. Just because the market price for carbon is high does not mean it is not working. However, the sulphur emissions trading program in the US has near legendary status among some in the environmental community because it was perceived to reduce the cost of abatement by an order of magnitude. In this case economists have showed that while there were likely gains due to use of the cap and trade system, the claim of an order of magnitude reduction in cost focused on some likely exaggerated early cost projections and some fortunate circumstances unrelated to emissions trading *per se* (e.g., deregulation of railroads that reduced the cost of transporting low sulphur coal from the Western US) (Ellerman *et al.*, 2000).

So far the experience with carbon trading in Europe is exactly the opposite of that with sulphur trading in the US. The permit trading price is an order of magnitude higher than what was expected. This would seem to create the risk of a perception that emissions trading has failed, and leads to excessively costly abatement. This would be an unfortunate and probably unwarranted conclusion. Just as casual observers of the sulphur market credited to emissions trading what were bad early estimates and lucky coincidence, the surprisingly high cost of carbon permits in the ETS may reflect overly optimistic initial estimates and unlucky coincidence. Investigating the surprising divergence between expectation and (early) reality is thus important, and this paper is a very first attempt.

In that regard, unlucky coincidence does appear to be an important explanation for higher prices than models had projected. Large increases in natural gas prices likely led to utilities relying more heavily on coal for generation than they otherwise would have, and made abatement through switching to gas an expensive option. At the same time, reduced hydro and possible concerns about nuclear electricity production likely had an effect. This could by our estimate explain a price of 2 to 5 but not 20 to 25 €/tCO_2.

The carbon market at this point is subject to very different expectations than was the sulphur market when it was established in the US. It is probably fair to say that the expectation in the sulphur market was that the cap established at the time was the ultimate cap. In contrast, in the carbon market, there is widespread recognition that the modest reductions in the ETS are part of an early test, and that caps will need to be tightened further in the future. The EU is bound by the Kyoto Protocol to make deeper cuts in the future, and the UK and France have set even more ambitious long-term reduction targets. If unused allowances could be banked, then the supply-demand situation in 2005 to 2007 would poorly predict prices because we would expect many firms to over comply and hold allowances for 2008–2012 or subsequent period when caps would tighten and prices would be higher. The hitch here is that there is no provision in the Kyoto Protocol that would allow banking into the first commitment period from some other system, and most of the EU NAPs specifically indicate banking is not allowed following guidance from the EC. Our analysis suggests that to generate prices that could support the current market price on the basis of banking would require the relatively extreme assumptions that during the Kyoto period the ETS would be extended to other sectors in Europe, but their would be no crediting of Joint Implementation, CDM, trades from other regions, or from non-CO_2 greenhouse gas abatement. It would also assume that firms were essentially certain that banking would be allowed even though, at this point, there would seem to be little reason to believe that it will and thus most banked allowances could be rendered worthless.

Another way in which future programs could affect current prices is if there is an expectation that future allocations of allowances will be based on actual emissions levels in 2005 to 2007. If this were to happen it would be a condition that would greatly concern economists because this would make the trading system work inefficiently. Essentially firms would not want to abate, expecting that high emissions would be rewarded with a high level of allowances for 2008 to 2012. We show through a simple example that this could have strong effects on prices, but to fully evaluate it would require deeper analysis than we could conduct at this point.

Another reason that expectations and model projections for the price may have been too low is that the model used here does not include transactions costs or may simply represent abatement as easier than it is in reality. There are many reasons, in addition to those already discussed, why a model might fail to reproduce the actual emission permit prices. For example, in the EPPA model there is perfect information about all markets, no monopoly power, and no government regulatory constraints on adjustments. Markets, including all factor markets, always clear immediately in the model, so there is never any unemployment or unused capacity. Output, input, and investment decisions are always just right. All of the conditions above might contribute to making the model's estimates of carbon prices lower than those that currently prevail in the ETS. Compelling quantitative analysis of these factors is difficult. Still, it is hard to reconcile the very wide difference if indeed the required reduction is on the order of 1 percent.

As economists we have a fair amount of faith in markets, and in the end models like those we have created supposedly are designed to represent market behavior. Thus, if the permit price response to the ETS remains similar to the early experience more work will be needed to reconsider model structure and the causes behind the divergence between simulated prices and reality. At this point many observers hold the view that the early market experience may not represent well the ultimate results for the three year ETS period. The market is just beginning, registries in some countries are still not operating, the first real reporting period is still several months away, and the shocks of rising gas prices and low hydro capacity have no doubt jolted firms under the ETS. The volume of trade thus far has been quite low relative to the total level of allocated allowances. Thus, there is reason to be cautious about reading too much into the early market price, and the jolts that have been experienced would be expected to push prices toward the high side. A few skittish firms could be pushing up prices on a relatively small volume of permits, while the more knowledgeable and cautious firms are waiting until they at least see results from the first year operation of the system, knowing that they can cover their emissions in that year by borrowing against the second year allocation.

The experiment with carbon trading in the European Union is important. The experience in terms of the market clearing price has been a surprise (if not a shock) based on expectations that the reductions required would be very mild. The high prices may mean that we need to reconsider the models we have used to estimate abatement costs, but unexpected shocks or expectations about the future may be strongly affecting the current market price. There are multiple real factors that may be contributing to these higher than expected prices. None of them on their own seem sufficient to explain the current prices. With over two years to go before the test phase of the ETS is complete, it is too early to make firm conclusions but it will be important to continue to monitor and evaluate the performance of the system because perceptions of its performance could well determine whether a greenhouse emissions trading system is expanded into a broader global system or not.

NOTES

* We gratefully acknowledge the helpful comments from Denny Ellerman, Richard Eckaus, and John Parsons, and the research assistance of Kelvin See. The CGE model underlying the analysis is supported by the US Department of Energy, US Environmental Protection Agency, US National Science Foundation, US National Aeronautics and Space Administration, US National Oceanographic and Atmospheric Administration; and the Industry and Foundation Sponsors of the MIT Joint Program on the Science and Policy of Global Change.
1. Estimates as published were in 1990 US$/tC, converted here to current (2005) $/tCO$_2$ and to €/CO$_2$ using the US implicit price deflator and the current $/€ exchange rate. The highest reported price estimate was about 100 €/tCO$_2$, considerably above the next highest at about 60 €/tCO$_2$.

REFERENCES

Babiker, M.H., J.M. Reilly, M. Mayer, R.S. Eckaus, I.S. Wing and R.C. Hyman (2001), 'The MIT Emissions Prediction and Policy Analysis (EPPA) Model: Revisions, Sensitivities, and Comparisons of Results', MIT Joint Program on the Science and Policy of Global Change, Report 71, Cambridge, MA.

Betz, R., W. Eichhammer and J. Schleich (2004), 'National allocation plans: analysis of the outcome of national negotiation processes', *Energy and Environment*, **15** (3), 375-425.

EC (2005), 'EU emission trading. An open scheme promoting global innovation to combat climate change', Brussels: European Commission.

EC (2003), 'Directive 2003/87/EC Establishing a Scheme for Greenhouse Emission Allowance Trading within the Community and amending Council Directive 96/61/EC', Brussels: European Commission.

EEA (2005), 'Annual European Community Greenhouse Inventory 1990–2003 and Inventory Report 2005', European Environment Agency Technical Report No 4/2005.

Ellerman, A.D., P. Joskow, R. Schmalensee, J.P. Montero and E.M. Bailey (2000), *Markets for Clean Air*, Cambridge UK: Cambridge University Press.

Paltsev, S., J.M. Reilly, H. Jacoby, R.S. Eckaus, J. McFarland, M. Sarofim, M. Asadoorian and M. Babiker (2005), 'The MIT Emissions Prediction and Policy Analysis (EPPA) Model: Version 4', MIT Joint Program on the Science and Policy of Global Change, Report 125, Cambridge, MA.

Paltsev, S., J.M. Reilly, H. Jacoby and K.H. Tay (2004), 'The Cost of Kyoto Protocol Targets: The Case of Japan', MIT Joint Program on the Science and Policy of Global Change, Report 112, Cambridge, MA.

Pew Center (2005), 'The European Union Emissions Trading Scheme (EU-ETS): Insights and Opportunities', Alexandria, VA: Pew Center.

Point Carbon (2005), 'Carbon Market Europe, September 23, 2005', A Point Carbon Publication, available at: http://www.pointcarbon.com

Reilly, J.M. and S. Paltsev (2005), 'An Analysis of the European Emission Trading Scheme', MIT Joint Program on the Science and Policy of Global Change, Report 127, Cambridge, MA.

See, C.S. (2005), 'Carbon Permit Prices in the European Emissions Trading System: A Stochastic Analysis', Master's Thesis, Technology and Policy Program, Massachusetts Institute of Technology, Cambridge, MA (June).

Sokolov, A., C. Schlosser, S. Dutkiewicz, S. Paltsev, D. Kicklighter, H. Jacoby, R.G. Prinn, C. Forest, J. Reilly, C. Wang, B. Felzer, M. Sarofim, J. Scott, P. Stone, J. Melillo and J. Cohen (2005), 'The MIT Integrated Global System Model (IGSM) Version 2: Model Description and Baseline Evaluation', MIT Joint Program on the Science and Policy of Global Change, Report 124, Cambridge, MA.

Weyant, J. and J. Hill (1999), 'Introduction and overview', *The Energy Journal*, Special Issue, The Costs of the Kyoto Protocol: A Multi-Model Evaluation, vii-xliv.

Zetterberg, L., K. Nilsson, M. Ahman, A.S. Kumlin and L. Birgersdotter (2004), 'Analysis of national allocation plans for the EU ETS', IVL Swedish Environment Research Institute, Report B1591, August.

5. Harmonizing emission allocation. What are the equity consequences for the sectors in and outside the EU-trading scheme

Tim Hoffmann, Andreas Löschel and Ulf Moslener

INTRODUCTION

Within the Burden Sharing Agreement (BSA), the European Union countries have agreed to reduce carbon dioxide emissions by individual amounts relative to their emissions in 1990 to reach the Kyoto target for the EU as a whole. Compared to today's emissions, the reduction requirements formulated within the BSA differ substantially among EU countries. Our analysis looks at the consequences this inequality – inherent to the Burden Sharing Agreement as it exists – may have on climate policy and compliance costs of individual EU Member States, in particular on the optimal design of their National Allocation Plans (NAPs) to implement the European Emissions Trading Scheme (ETS).

We demonstrate that attempts to harmonize the free allocation of emission permits across countries may introduce significant efficiency costs. It dramatically increases the dispersion of average costs per ton of reduced carbon dioxide within the sectors that are not subject to the emissions trading scheme. In some cases, the average cost per ton of abated CO_2 increases substantially, such that for these countries the use of flexible mechanisms of the Kyoto Protocol (such as JI and CDM) seems inevitable to meet the target.

The following section will briefly introduce the European Emissions Trading Scheme the problem of formulating National Allocation Plans in general and the political motivation to harmonize allocation. This is followed by an examination of how harmonizing the allocation can generate efficiency costs and thereby undermine the instrument of permit trading. Furthermore, harmonizing the allocation to the sectors within the trading regime will have consequences for the parts of the economy that are not eligible for trading. These issues will then be illustrated by a simple partial equilibrium model for the former EU-15.

DESIGNING NATIONAL ALLOCATION PLANS

The key instrument for climate policy in the EU is the Emissions Trading Scheme, which is in force since the beginning of 2005 (EU, 2003a, 2004). It requires the energy-intensive sectors (mainly industry and the power sector) to take part in a regulatory regime of EU-wide tradable permits. Within this scheme, a number of important decisions have been left to the member state authorities, particularly the right to formulate the so-called National Allocation Plans.

In these NAPs, the national regulators distribute the national emission budgets to the different sectors of the economy. They include the sectors which will be part of the emissions trading scheme as well as the sectors not covered by the directive. In other words, the NAPs determine not only how many emissions the different sectors (trading emissions or not) will get but also how many emission permits will exist on the whole market until 2007 – the end of the first period of the EU ETS. Thereafter, new NAPs will determine the permit endowment for the first phase of the Kyoto Protocol – when the reduction commitments of the members to the protocol will finally have to be met according to international law. The design of these allocation plans to come into force in 2008 will show how the different reduction requirements resulting from the EU BSA are shared between the sectors subject to emissions trading, the other sectors which do not trade emissions, or the state (e.g., by buying credits using the flexible instruments of the Kyoto Protocol).

The European Commission laid out several criteria to accept the allocation plans. *Inter alia*, (i) they should be in line with the commitments under the BSA, (ii) they should not be abused to provide some kind of state aid, and (iii) Member States are obliged to distribute the allowances mainly for free, i.e., the maximum fraction to be auctioned should not exceed 10 percent of a country's permits.

The challenges posed by these criteria in combination with the goal of the directive itself, namely the cost efficient reduction of CO_2 emissions, have been examined previously (see Böhringer and Lange, 2005 or Böhringer *et al.*,

2005). The efficiency costs are closely related to the fact that the ETS introduces two separate areas of regulation: the sectors subject to emissions trading and the rest of the emission sources where other regulation on the member state level has to be applied. Given the fixed but differentiated commitments under the EU BSA, allocating the efficient emission budgets to the sectors not subject to the directive will leave the sectors that are trading emissions with very different shares of their previous emissions allocated for free. This will potentially lead to competitive disadvantages of emissions trading sectors in countries where the national emissions budget is comparatively small compared to their actual emissions.

Thus, in addition to the political pressure that usually comes with the introduction of a new regulation scheme, the states are faced with the questions of how they distribute permits to the different sectors trying to minimize their nationwide (and also European) compliance cost on the one hand, and how much they take into account the potential disadvantages that their sectors – within as well as out of the scheme – may suffer if their relative reduction requirements exceed those of their competitors in and outside of Europe. It should be noted that allowing for an auction of a larger fraction of the permits would contribute to solving the problem, at least with respect to competitors within the European Union. The debate on harmonizing the allocation process is related to these concerns about competitiveness.

The cost burden which may lead to competitive disadvantages of a sector is determined by the sectoral emissions budget and by the set of (cheap) abatement options (i.e., the shape of the marginal abatement cost function), but also by the overall emission intensity of the sector. Abstracting from the latter two, i.e., considering only the reduction requirement, some request the harmonization of the so-called fulfillment factor. This is the amount of emissions that are allocated to the sector (or firm) divided by its actual emissions.

Table 5.1 contains a condensed characterization of National Allocation Plans across the EU-15 Member States mainly based on the most recent screening by Gilbert *et al.* (2004). The column labeled '$\bar{E}_r$' in the table indicates the overall emission budget for each Member State r, as follows from the EU BSA. The subsequent column provides the emission cap $\bar{E}_r^{DIR}$ for sectors (*DIR*) that are covered by the EU Directive (reflecting the NAP, based on Gilbert *et al.*, 2004). The residual $\bar{E}_r - \bar{E}_r^{DIR}$ yields the implicit emission constraint $\bar{E}_r^{NDIR}$ for sectors (*NDIR*) that are not covered by the emissions trading system. Depending on the projected baseline emissions for *DIR* and *NDIR* sectors in our reference year ($\bar{E}_{r,2005}^{DIR}$ and $\bar{E}_{r,2005}^{NDIR}$, based on EU, 2003b), the final columns of the table report (i) the so-called fulfillment factor

$\lambda_r = \bar{E}_r^{DIR}/ \bar{E}_{r,2005}^{DIR}$, i.e., the fraction of baseline emissions that are freely allocated to the *DIR* sectors as allowances and (ii) the implicit relative reduction requirements for *NDIR* sectors $\mu_r = \bar{E}_r^{NDIR}/ \bar{E}_{r,2005}^{NDIR}$. We see that fulfillment factors vary very little as compared to the implicit relative reduction requirement in the *NDIR* sectors. Member States appear to have chosen comparatively high and somehow harmonized fulfillment factors for the *DIR* sectors. In this chapter, we examine the effect of harmonizing the fulfillment factor and make a first attempt to illustrate what can be expected if the different shapes of abatement cost functions are taken into account.

Table 5.1   Segmentation of emission budgets under national allocation plans for EU-15

	$\bar{E}_r$	$\bar{E}_r^{DIR}$	$\bar{E}_r^{NDIR}$ $= \bar{E}_r - \bar{E}_r^{DIR}$	$\bar{E}_{r,2005}^{DIR}$	$\bar{E}_{r,2005}^{NDIR}$	$\lambda_r =$ $\bar{E}_r^{DIR}/\bar{E}_{r,2005}^{DIR}$	$\mu_r =$ $\bar{E}_r^{NDIR}/\bar{E}_{r,2005}^{NDIR}$
Austria	47.9	22.1	25.8	23.52	36.78	0.940	0.702
Belgium	98.3	61.6	36.8	59.07	54.52	1.042	0.675
Denmark	41.7	20.6	21.1	24.20	24.20	0.850	0.872
Finland	53.2	33.7	19.5	34.35	21.05	0.980	0.927
France	354.1	85.3	268.8	85.78	304.11	0.995	0.884
Germany	745.0	481.2	263.8	481.20	334.40	1.000	0.789
Greece	88.9	58.7	30.2	58.68	39.12	1.000	0.772
Ireland	33.6	13.8	19.7	14.27	30.32	0.970	0.650
Italy	365.4	228.2	137.2	212.52	204.20	1.074	0.672
Netherlands	143.7	89.9	53.9	87.24	77.37	1.030	0.697
Portugal	49.5	27.2	22.3	26.32	34.87	1.035	0.640
Spain	234.4	111.4	123.0	118.50	174.09	0.940	0.707
Sweden	52.6	15.3	37.4	15.25	37.36	1.000	1.001
UK	498.0	240.6	257.4	242.37	284.53	0.993	0.905
EU–15 (total)	2806.3	1489.5	1316.8	1483.27	1656.93	1.004	0.795

Source: Gilbert *et al.,* 2004

Key: $\bar{E}_r$:   total emission budget under the EU Burden Sharing Agreement

$\bar{E}_r^{DIR}$:   emission budget of DIR sectors under the National Allocation Plans

$\bar{E}_r^{NDIR}$: emission budget of NDIR sectors under the National Allocation Plans

$\bar{E}_{r,2005}^{DIR}$: projected business-as-usual emissions for the DIR sectors in 2005

λ_r: fulfillment factor

μ_r: relative reduction target of NDIR sectors.

HARMONIZING THE FULFILLMENT FACTOR

In principle, harmonized fulfillment factors can be achieved in different ways. Given that auctioning larger fractions of the emission budget will be permitted by the directive, harmonization would even be possible while preserving efficiency (see Böhringer and Lange, 2005). The Member States could allocate whatever matches the efficient level to their sectors not subject to emissions trading. Then Member States could agree on a harmonized – comparatively small – fulfillment factor for the emissions trading sectors. Each member state could then auction off his remaining emission permits.

But to date auctioning the permits is restricted by the directive. If they want to harmonize the fulfillment factors, Member States are forced to move away from the country-specific – but efficient – allocation to the sectors that cannot trade emissions. Inefficiencies are now predominantly introduced in these non-trading sectors. The trading sectors may have an overall emission target that might not be efficient, but through trade they are still able to implement this target at (second-best) minimum costs. Therefore, using an identical fulfillment factor comes at economic costs as compared to an efficient allocation between the sectors in and out of the trading scheme. The costs of implementing the emission targets will depend on the level of the harmonized fulfillment factor and there will be a cost minimizing *unified* fulfillment factor, but it cannot lead to efficient implementation of the Kyoto targets at the European level. In fact, efficiency costs are likely to be substantial (Böhringer *et al.*, 2005).

We now take a closer look at the consequences of harmonizing the fulfillment factor for the sectors out of the emissions trading scheme. In the efficient case (with diverse fulfillment factors) the marginal abatement costs are equalized across countries and sectors. Differences in the efficient relative reduction requirements for the sectors out of the trading scheme from country to country reflect the different shapes of the marginal abatement costs. If the fulfillment factor of the trading sectors is harmonized, the relative reduction requirements of the other sectors are directly affected *via* the given national BSA emission budgets of the individual Member States. This may alter the spread of the reduction requirements across countries, and it may increase the spread of average compliance cost per ton of (reduced) carbon, thereby to some extent shifting the problem that the harmonization is trying to avoid in the trading sectors towards the sectors outside the scheme.

It is subject to a quantitative assessment whether this problem is substantial in the sectors that do not trade their emissions. First, some of the sectors not included in the trading regime are less emission intensive, e.g., households or services. Regulation such as emission taxes in these sectors – if differing across countries – might simply not matter that much if emissions play a minor role. Second, the exposure to the world (or European) market may be different among the emissions trading and non-trading parts of the economy. Less exposure to international markets (e.g., local services) can help prevent consumers from switching to products from other, unregulated firms.

Furthermore, it is possible that national governments take on some of the cost burden of reducing emissions by buying permits generated from the flexible mechanisms of the Kyoto Protocol. Of course, the government would have to finance this and somehow raise the money, which may well result in another burden on the economic sectors. But the reduction commitment can in principle be met by using the state budget.

This completes the picture as to where the diverging relative reduction targets of the Member States – that were once fixed in the BSA – may be accounted for. The differences can be distributed among the different sectors (especially the trading and the non-trading sectors) but they can also appear in the individual Member States' budget. The consequences for the overall efficiency will vary depending on the way the differences are distributed and on the economic instruments that are used.

NUMERICAL ANALYSIS OF HARMONIZATION

Non-technical model description

We will illustrate the problem of harmonizing the fulfillment factor by way of a simple partial equilibrium model based on marginal abatement cost curves for the former EU-15.[1] The model considers 15 regions (countries) where each region is constrained by an overall emission budget (as given by the EU Burden Sharing Agreement). In designing a National Allocation Plan, a Member State has to consider the abatement costs of the sectors covered by the emissions trading directive (*DIR*) and the sectors *not* covered (*NDIR*). We make the standard assumption that abatement costs in each region and each sector can be characterized by decreasing, convex and differentiable functions of the abatement level.

With international emissions trading, efficient national regulation comes down to minimizing compliance costs as the sum of abatement costs across all domestic sectors and the costs of buying emission allowances from the international market at an endogenous price.[2] Differentiating the minimization problem with respect to both *DIR* and *NDIR* emission levels, the associated first-order conditions state that marginal abatement costs are equalized across all sectors at the international emissions price. Optimal emission levels for each region follow as the aggregate optimal emission levels of *DIR* and *NDIR* sectors in the respective region. The difference between the exogenous total emission budget and aggregate optimal emissions yields the optimal trade volume in emission allowances. The efficient allocation, or fulfillment factor λ, which reports the fraction of business-as-usual emissions that are freely allocated to *DIR* sectors as allowances, is determined implicitly as well as the implicit relative reduction requirement μ for *NDIR* sectors, which reports the fraction of their business-as-usual emissions that has to be reached by complementing domestic abatement policies.

A detailed algebraic description of the model can be found in Böhringer *et al.*, 2005. Numerically, our model is implemented as a Mixed Complementarity Problem (MCP) formulated in GAMS (Brooke *et al.*, 1987) using PATH (Dirkse and Ferris, 1995) as a solver. The GAMS file and the EXCEL reporting sheet can be downloaded from the web-site: http://brw.zew.de/simac/model.zip.

Model parameterization

Marginal costs of emission abatement may vary considerably across countries and sectors due to differences in carbon intensity, initial energy price levels, or the available carbon substitution possibilities. Continuous marginal abatement cost curves for the *DIR* and *NDIR* sectors in EU countries can be derived from a sufficiently large number of discrete observations for marginal abatement costs and the associated emission reductions in the *DIR* and *NDIR* sectors. In applied research, these values are often generated by partial equilibrium models of the energy system (such as the POLES model by Criqui and Mima (2001) or the PRIMES model by Capros *et al.*, 1998) that embody a detailed bottom-up description of technological options. Another possibility is to derive marginal abatement cost curves from computable general equilibrium (CGE) models (see, e.g., Eyckmans *et al.*, 2001). We adopt the latter approach and generate a reduced form of complex CGE interactions in terms of marginal abatement cost curves.

In order to obtain such marginal abatement cost curves for the *DIR* and *NDIR* sectors across EU countries, we use the PACE model – a standard multi-region, multi-sector CGE model for the EU economy (for a detailed algebraic exposition, see Böhringer, 2002) which is based on recent consistent accounts of EU Member States' production and consumption, bilateral trade and energy flows for 1997 (as provided by the GTAP5-E database – see Dimaranan and McDougall, 2002). With respect to the analysis of carbon abatement policies, the sectors in the model have been carefully selected to keep the most carbon-intensive sectors in the available data as separate as possible. The energy goods identified in the model include primary carriers (coal, natural gas, crude oil) and secondary energy carriers (refined oil products and electricity). Furthermore, the model features three additional energy-intensive non-energy sectors (iron and steel; paper, pulp and printing; non-ferrous metals) whose installations – in addition to the secondary energy branches (refined oil products and electricity) – are subject to the EU emissions trading system. The remaining manufacturers and services are aggregated to a composite industry that produces a non-energy-intensive macro good, which together with final demand captures the activities (*NDIR* segments) that are not included in the EU trading system.

To generate our reduced form model, we perform a sequence of carbon tax scenarios for each region where we impose uniform carbon taxes (starting from 0 € to 200 € per ton of carbon in steps of 1 €). We thereby generate a large number of marginal abatement costs, i.e., carbon taxes, and the associated emission reductions in *DIR* and *NDIR* sectors. We then apply a least-square fit by a third-degree polynomial. This approach is, however, neglecting market interaction and spillover effects.[3] Apart from terms-of-trade effects, other potentially important general equilibrium interactions concern revenue-recycling. It is well-known that the manner in which revenues from environmental regulation are recycled to the economy can have a larger impact on the gross costs of environmental policy (see Goulder, 1995, or Bovenberg, 1999).

Likewise, in case the Member States' governments intend to buy larger amounts of emission permits via the flexible mechanisms of the Kyoto Protocol, the way in which the governments raise the money may play a more important role.

Simulation results: Efficient versus harmonized fulfillment factors

First we demonstrate the efficiency losses that result from a harmonization of the fulfillment factor for *DIR* sectors. Figure 5.1 displays the total compliance costs (C_{Total}) for the whole EU-15 as well as the compliance costs of the *DIR* (C_{DIR}) and the *NDIR* (C_{NDIR}) sectors if fulfillment factors (λ) are harmonized at different levels (within a range from 0.5 to 1.1). It is obvious that the cost burden of the *DIR* sectors decreases in rising fulfillment factors since higher λ directly translate into reduced emission abatement obligations. Given an overall emission constraint, a reduction of relative abatement requirements in the *DIR* sectors inevitably shifts the burden towards the *NDIR* sectors where increasing abatement costs can be observed.

Total compliance costs reach an overall cost minimum at a harmonized λ of 0.8. Figure 5.1 also depicts the total compliance costs (C_{opt}) for the case of efficient but country-specific (non-harmonized) fulfillment factors. In this illustration, an efficient distribution of country-specific emission budgets across *DIR* and *NDIR* sectors would reduce overall compliance costs to less than 50 percent as compared to the harmonized case. Furthermore, the additional costs of harmonization are distributed across *DIR* and *NDIR* sectors. While harmonization increases costs in *DIR* sectors by approximately 50 percent, additional costs for the *NDIR* sectors are more than five times higher than in the efficient case.

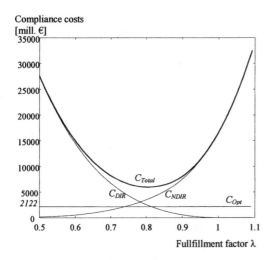

Compliance costs
[mill. €]

Fullfillment factor λ

Figure 5.1   Compliance costs for the EU-15 for efficient but country-specific fulfillment factors (horizontal line) and for harmonized factors (u-shaped curve)

It is also evident that a harmonization of fulfillment factors for *DIR* sectors affects relative reduction requirements of the *NDIR* sectors (μ). Figure 5.2 indicates the correlation between the relative reduction targets of *DIR* and *NDIR* sectors (λ and μ) in each Member State and the corresponding average compliance costs per ton of CO_2 in the respective sectors and countries. By definition, harmonizing the fulfillment factors of *DIR* sectors eliminates their variance compared to the efficient case shown in the lower part of Figure 5.2 where country-specific fulfillment factors for *DIR* sectors range from 0.29 up to 1.09. Given this spread, it is not surprising that there are concerns in some sectors that they might be disadvantaged relative to their competitors in other EU Member States. Relative reduction targets of *NDIR* sectors vary between 0.75 and 1.39. Surprisingly, the overall reduction requirements of the Member States as stated in the BSA vary a lot less: if compared to the projected emissions in 2005 (EU, 2003b), the available relative budget for the first

Kyoto phase ranges from approximately 1 (Sweden) to about 0.75 (Ireland). In the efficient case, average compliance costs of the *DIR* sectors – composed of sectoral abatement costs and costs or revenues for buying or selling permits on the international market – range from - 25 up to almost 27 €/tCO$_2$. Average compliance costs of the *NDIR* sectors – composed only of abatement costs – range from less than 1 to 10 €/tCO$_2$.

As can be seen in the upper part of Figure 5.2, applying a harmonized λ of 0.8 does not substantially inflate the variance of the relative reduction requirements of the *NDIR* sectors (between 0.77 and 1.49). But this harmonization has a strong effect on the variance of the average compliance cost per ton of CO$_2$.

While the spread of the *DIR* sectors' average abatement costs slightly decreases and these abatement costs have an upper bound on their average compliance cost per ton of about 11 €, the average costs for the individual *NDIR* sectors may reach more than 130 €/tCO$_2$ for harmonization at λ=0.8. This spread increases even more when a higher harmonized λ is chosen. The middle part of Figure 5.2 depicts the results for a harmonized λ of 0.95 which reflects an over-allocation of *DIR* sectors in the EU-15. This situation can be considered similar to the allocation plans that have actually been designed by the Member States. Not surprisingly, compliance now becomes cheap for *DIR* sectors but the majority of *NDIR* sectors face substantially higher average compliance costs up to more than 200 Euros per ton of CO$_2$. This comes with a shift of the relative reduction requirements in the *NDIR* parts towards more severe targets.

On the whole, the harmonization of the assignment in the *DIR* sectors does not seem to substantially inflate the variance of the relative emission assignments to the *NDIR* sectors. But it increases its average costs per ton of reduced emissions significantly – in some cases drastically.

Given that an (at least partial) harmonization seems to be the alternative that has been chosen in the political process, it seems obvious that this would come with significantly stricter requirements in the *NDIR* sectors – especially in terms of average cost per abated ton of CO$_2$ – of some countries. In these cases, it can be expected that the emission target will have to be implemented by adding the flexible mechanisms of the Kyoto Protocol to the domestic sectoral efforts. Some of the costs may then be borne by the state budget.

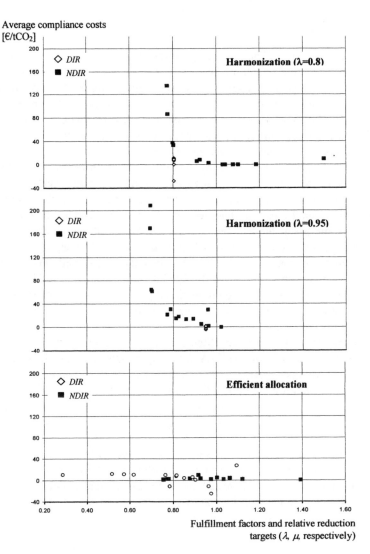

Note: the three scenarios are (i) the harmonized fulfillment factor at the second best level (λ=0.8), (ii) a harmonized factor which leads to overallocation (λ=0.95), and (iii) the efficient allocation.

Figure 5.2   Average compliance cost of individual countries in the DIR and NDIR sectors and their fulfillment factor

CONCLUSIONS

We analyze the effects of harmonizing the fulfillment factors of the sectors included in the trading regime of the EU. The most significant result is that harmonization under the auctioning restrictions – although sometimes politically desirable – will come at high efficiency costs. However, when striving for cost efficient emission reduction, the abatement requirements for the different countries – in the sectors within as well as outside the directive – vary a lot, even if compared to the national reduction requirements as laid down in the European burden sharing agreement. Harmonization eliminates this variance in the *DIR* sectors and seems not to inflate it significantly in the other sectors. The larger fraction of the additional cost burden from harmonization can be expected to rest on the sectors outside the trading regime. In some cases, the average cost per abated ton of carbon dioxide is increased to unrealistic levels if strict harmonization and meeting the Kyoto target without the use of its flexible instruments is assumed. For these countries, one may expect the use of flexible instruments which could involve the Member States as actors in the international permit market beyond the EU, and it may at the same time mean that the Member States' budget could pay for some of the compliance costs.

As for the inequalities between Member States in terms of the reduction requirements as laid down in the BSA, it largely remains a political question to wich extent Member States decide to harmonize the allocation process for the first Kyoto commitment period from 2008 to 2012. It should be kept in mind that the relative emission requirement alone may not be a good measure for the influence of the allocation on the competitive disadvantages resulting from it. The same holds for the average compliance cost per ton of abated CO_2. Our analysis at least accounts for different shapes of the abatement cost functions. For future analysis, aspects like the emission intensity of the output value, which is inter alia influenced by historical climate policies, or the exposure of sectors to the world market would have to be considered as well.

NOTES

*   We are grateful to the Research Group in Economics, Energy and the Environment (rede) at the University of Vigo for organizing the inspiring First Atlantic Workshop on Energy and Environmental Economics in A Toxa.

1.   The interested reader can access the model via a web-based interface (http://brw.zew.de/simac/).

2. The cost-efficient allocation depends on whether the EU system is closed to the world market or it is possible to use credits obtained from CDM projects or other countries. Although our model incorporates a world market, the present analysis will abstain from explicitly considering the governments as actors on permit markets such that for the illustrative purposes of distributional and efficiency aspects, a partial model seems justified.

3. For analyses illustrating the potential importance of the effects, see Böhringer (2002); Böhringer and Rutherford (2002); Bernard *et al.* (2003); and Klepper and Peterson (2002).

REFERENCES

Bernard, A., S. Paltsev, J.M. Reilly, M. Vielle and L. Viguier (2003), 'Russia's role in the Kyoto Protocol', MIT Joint Program on the Science and Policy of Global Change, Report No. 98. Cambridge, MA.

Böhringer, C. (2002), 'Industry-level emission trading between power producers in the EU', *Applied Economics*, **34** (4), 523-533.

Böhringer, C. and A. Lange (2005), 'Mission Impossible!? – on the harmonization of national allocation plans under the EU emissions trading directive', *Journal of Regulatory Economics*, **27** (1), 51-71.

Böhringer, C. and T.F. Rutherford (2002), 'Carbon abatement and international spillovers', *Environmental and Resource Economics*, **22** (3), 391-417.

Böhringer, C., T. Hoffmann, A. Lange, A. Löschel and U. Moslener (2005), 'Assessing emission allocation in Europe. An interactive simulation approach', *The Energy Journal*, **26** (4), 1-21.

Bovenberg, A.L. (1999), 'Green tax reforms and the double dividend. An updated reader's guide', *International Tax and Public Finance*, **6**, 421-443.

Brooke, A., D. Kendrick and A. Meeraus (1987), *GAMS. A User's Guide*, San Francisco: Scientific Press.

Capros, P., L. Mantzos, D. Kolokotsas, N. Ioannou, T. Georgakopoulos, A. Filippopoulitis and Y. Antoniou (1998), 'The PRIMES Energy System Model – Reference Manual', National Technical University of Athens. Document as peer reviewed by the European Commission, Directorate General for Research.

Criqui, P. and S. Mima (2001), 'The European greenhouse gas tradable emission permit system. Some policy issues identified with the POLES-ASPEN model', *ENER Bulletin*, **23**, 51-55.

Dimaranan, B. and R.A. McDougall (2002), *Global Trade, Assistance and Production. The GTAP 5 Data Base*. West Lafayette, Center for Global Trade Analysis. Purdue University.

Dirkse, S. and M. Ferris (1995), 'The PATH solver, A non-monotone stabilization scheme for mixed complementarity problems', *Optimization Methods and Software*, **5**, 123-156.

EU (2003a), Directive Establishing a Scheme for Greenhouse Gas Emission Allowance Trading within the Community and Amending Council Directive 96/61/EC. European Commission, Brussels. Available at: http://europa.eu.int/eur-lex/fr/com/ pdf/2001/ fr_501PC0581.pdf.

EU (2003b), *European Energy and Transport Trends to 2030*, European Commission, Brussels.

EU (2004), Directive 2004/101/EC, amending Directive 2003/87/EC establishing a scheme for greenhouse gas emission allowance trading within the Community, in respect of the Kyoto Protocol's project mechanisms. European Commission, Brussels. Available at: http://europa.eu.int/comm/envoirnment/climat/emission/pdf/ dir_2004_101_en.pdf.

Eyckmans, J., D. van Regemorter and V. van Steenberghe (2001), 'Is Kyoto fatally flawed? – An analysis with MacGEM', Working Paper Series – Faculty of Economics, No. 2001-18, Catholic University of Leuven, Leuven.

Gilbert, A., J.W. Bode and D. Phylipsen (2004), 'Analysis of the national allocation plans for the eu emissions trading scheme', Ecofys Interim Report, Utrecht. Available at: http://www.ecofys.com/com/publications/documents/Interim_Report_ NAP_Evaluation_180804.pdf.

Goulder, L.H. (1995), 'Environmental taxation and the double dividend. A reader's guide', *International Tax and Public Finance*, **2**, 157-183.

Klepper, G. and S. Peterson (2002), 'On the robustness of marginal abatement cost curves. The influence of world energy prices', Kiel Working Paper No. 1138. Institute for World Economics, Kiel.

6.   The effects of a sudden CO_2 reduction in Spain

Xavier Labandeira and Miguel Rodríguez

INTRODUCTION

The phenomenon of climate change, with a growing scientific consensus regarding causalities and associated damage, forms one of the most serious threats that humanity faces in the coming decades. The growing concentration of greenhouse-effect gases in the atmosphere is expected to provoke very significant physical and economic effects throughout the planet (elevations in sea levels, massive precipitation, serious droughts, etc.). In particular, carbon dioxide (CO_2) is the main cause of global warming, representing around 80 percent of total greenhouse gas precursors.

The seriousness of the problem has driven an important process of international agreement in the last few years over the control of these emissions, mainly through the Kyoto Protocol. This establishes the commitment of developed countries to reduce their greenhouse effect gases by an average of 5 percent with respect to 1990 as the year of reference. The refusal of the United States to ratify the Kyoto Protocol generated serious doubts regarding its final implementation, although the leading role of the European Union (EU) and other developed countries made this possible in February, 2005.

In fact, the Kyoto Protocol guidelines were accepted by the European Commission as early as in April 2002, laying down a distribution system of emission reduction efforts among the member countries in order to reach the objective of 8 percent for the entire EU. Among the measures adopted by the EU for the fulfilment of the Protocol, there is a Directive establishing a scheme for carbon dioxide emission allowance trading within the Community, which will be in operation as of 2005. The sphere of application to the market is limited, so only certain sectors will be regulated by this measure (electricity generation, petroleum refinement, the industries of iron and steel, cement, lime, glass, ceramics, brick and tile, paper and paper pulp). As a result, EU member countries have designed national plans for the allocation of rights among sectors. Within the basic principles of national allocation plans, the necessity for additional measures is also established to assure monitoring of sectors not included in the market, since they generate more than 50 percent of the CO_2 emissions. Such measures are oriented toward policies of energy saving and energy efficiency, although this does not rule out the use of environmental taxes.

With respect to the situation in Spain, the distribution of emissions was beneficial in permitting Spain to increase its emissions up to a maximum of 15 percent in the period of 2008–2012. However, the economic growth of recent years, together with the lack of political initiatives, has resulted in greatly increased energy consumption in Spain. Indeed in 2004, Spain's emissions of greenhouse gases had grown by approximately 40 percent with respect to 1990, an unsustainable performance from any point of view, be it political, economic or environmental.

In this context, corrective and stringent public policies are to be expected in the short term, and thus an insight into their effects seems particularly necessary. This is also necessary because Spain, given its geographical location, will probably suffer climate change more intensely than most other EU countries. In this chapter we analyze the effects of attaining different environmental objectives: from a reduction of 2 percent in the emissions of CO_2 (the annual reduction necessary, from 2004 onwards, in order to fulfil the objectives in the current Spanish national allocation plan) up to a maximum of 16 percent (the overall objective).

The results obtained show that the short-term cost of reducing Spanish emissions of greenhouse gases at the annual rate of 2 percent has little effect in terms of employment as well as in economic activity. However, a reduction of 16 percent in the short term would have a relatively considerable cost. We should conclude, therefore, that the inhibition of Spanish public policies towards fulfilling the Kyoto Protocol in recent years will increase the readjustment costs for the Spanish economy. Further delays in implementing the policy instruments for controlling Spanish emissions may mean major sacrifices in the future.

Our analysis of the effects of fulfilling the Kyoto Protocol in Spain is especially relevant owing to the scarce empirical evidence available. To our knowledge there are only two pieces of research that apply specifically to Spain, and these studies analyse the possible existence of double dividends of green tax reforms in Spain using static general equilibrium models (Manresa and Sancho, 2005; Faehn *et al.*, 2005).[1] However, the results are not fully satisfactory either because of a lack of substitution possibilities between energy goods and value added in the production function, of an inappropriate modelling of CO$_2$ emissions or due to the simulation of *ad-hoc* policies.

This chapter is structured into four sections, including this introduction. In section 2, the methodological approach is explained, with a description of the theoretical model and the empirical implementation. Section 3 presents the policies considered and the results obtained from those simulations. Finally, section 4 includes the main conclusions of the paper and some policy implications.

THE COMPUTABLE GENERAL EQUILIBRIUM MODEL

To evaluate the efficiency effects of environmental and energy policies, we use a multi-sectorial static applied general equilibrium (AGE) model for an open and small economy such as Spain.[2] This kind of model allows a greater breakdown of energy goods, an important feature of the model in order to take into account the heterogeneity of energy consumption between sectors. Therefore, the AGE can, to some extent, take into account, the different services provided by energies (intermediate inputs for production of electricity; lighting, heating and transport services for firms and institutions, etc.) and differences in CO$_2$ emission factors. We likewise include the conclusions of different chapters by Dean and Hoeller (1992), Clarke *et al.* (1996) and Repetto and Austin (1997). They highlight the importance of breaking down the energy assets in the economy so as not to produce biased estimation of the costs of environmental policies.

There are 17 productive sectors in the economy and therefore 17 commodities. Industries are modelled through a representative firm. They minimize costs subjected to null benefits at the equilibrium. Output prices are equal to average production costs, as we assume perfect competition and constant returns to scale. The production function is a succession of nested constant elasticity of substitution (CES) functions, as illustrated in Figure 6.1, that combine intermediate inputs, capital, labour and different energies. The energy goods are taken out from the set of intermediate inputs. They are included in a lower nest within the production function, allowing for more flexibility and substitution possibilities (from dirtier to cleaner energies on the basis of emission factors). Therefore, our model is similar, although with some changes, to that used by Böhringer *et al.* (1997).[3]

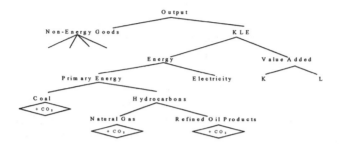

Source: authors' elaboration.

Figure 6.1   Production technology structure chain

We follow the Armington approach to model the international trade of goods as usual in the literature (Shoven and Whalley, 1992). First of all, imported products are imperfect substitutes for national production. Secondly, maximization of benefits by each sector, determined *via* a constant elasticity of transformation (CET) function, allocates the supply of goods and services between the export market and domestic consumption. Since the Spanish economy is small and most commodity trade is made with EMU countries, there is no exchange rate (it is fixed) and all agents face exogenous world prices.

Capital supply is inelastic and perfectly mobile between sectors, but immobile internationally. The model assumes a competitive labour market and therefore an economy without involuntary unemployment. The labour supply made by households to maximize utility is also perfectly mobile between sectors but immobile internationally.

The public sector collects direct taxes (income taxes from households, and wage taxes from households and sectors) and indirect taxes (from production and consumption). Endowment of capital for the government, transfers with other institutions and public deficit are exogenous variables. The consumption of goods and services by the government is determined by a Cobb-Douglas function. Therefore, total public expenditure, capital income and tax revenues have to be balanced in order to satisfy the budget restriction.

The representative household has a fixed endowment of time which is allocated between leisure and labour. It maximizes utility, which is a function of leisure and a composite good made up of goods and savings, subject to the budget constraint. It is assumed, as in Böhringer and Rutherford (1997), that consumers have a constant marginal propensity to save. Consumption of goods and services is defined by a nested CES function, as shown in Figure 6.2, with special attention being paid to the consumption of energy goods. An important feature of the AGE model is the distinction between energy for the house, energy for private transport and other products. Other non-energy goods are a composite good formulated via a Cobb-Douglas function.

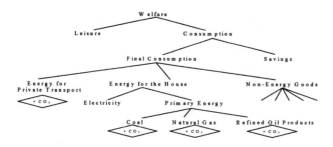

Source: authors' elaboration.

Figure 6.2   Chained household consumption function structure

The AGE model represents a structural model based on the Walrasian concept of equilibrium. Therefore, for each simulated policy, the model must find a set of prices and quantities in order to clear up all markets (capital, labour and commodities). Total savings in the economy is defined endogenously, and is equal to the sum of savings made by each of the institutions. The macroeconomic equilibrium of the model is determined by the exogenous financing capacity/need of the economy with the foreign sector, i.e. the difference between national savings, public deficit and national investment. The latter is a composite good by a Leontief function that incorporates the different commodities used in gross capital formation. International prices, transfers between the foreign sector and other institutions, and foreigners' consumption of goods and services when staying in Spain are exogenous variables. Therefore, exports and imports have to be balanced in order to satisfy the restriction faced by the foreign sector.

The model simulates energy-specific CO$_2$ emissions produced by different sectors and institutions. Therefore, we do not simulate the emissions made by some industrial production processes such as cement, chemical, etc. They only represented about 7 percent of total Spanish CO$_2$ emissions in 1995 (INE, 2002a). Emissions are assumed to be exclusively generated during the combustion processes of fossil fuels. Thus, there is a technological relationship between the consumption of fossil fuels in physical units and emissions for coal, refined oil products and natural gas, respectively.

The model database is a national accounting matrix for the Spanish economy (NAM-95), erected on the basis of the national accounts for 1995.[4] Furthermore, we have extended the database with environmental data from different statistical sources (INE, 2002a; IEA, 1998) relating consumption of different fossil fuels and emissions for each sector and institution. Based on the information obtained from the NAM-95, the model's parameters can be gauged by calibration:[5] tax rates or technical coefficients for production, consumption and utility functions. The criterion for calibrating the model is that the AGE model replicates the information contained in the NAM-95 as an optimum equilibrium, which will be used as a benchmark.[6] Certain parameters, such as elasticities of substitution, have not been calibrated, but taken from pre-existing literature (i.e., the wage elasticity of the labour supply).[7]

THE COSTS OF KYOTO FOR SPAIN

Simulated policies

The national plan for the allocation of emission rights, with final approval by the Spanish government in early 2005, establishes that between 2008 and 2012 the average amount of emissions should be below 124 percent of 1990 emissions. This value is the result of the sum of the maximum limit given by the EU for Spain (+15 percent), the estimation of the absorption of drains (-2 percent) and the emissions credits purchased in foreign markets (-7 percent). Assuming a 40 percent increase of 1990 emissions in 2004, it would therefore be necessary to reduce Spanish total emissions in 2012 by 16 percent or to reduce them by 2 percent every year.

In order to analyze the effects of fulfilling the Kyoto Protocol in Spain, we have considered different objectives: from a reduction policy of 2 percent in Spanish greenhouse gas emissions (equivalent to the necessary annual rate of reduction) to a drastic cutback equal to 16 percent (the overall objective in the Spanish national plan). Thus, we may explore the relationship between the intensity of the environmental objectives and their economic and social effects.

The policy instrument that we have considered is a tax on CO_2 emissions, with revenues returned to citizens through lump-sum transfers and subject to the restriction that public expenditure should remain constant in real terms. Therefore, in this chapter the environmental tax has the sole intention of reducing emissions of carbon dioxide. Indeed, the reimbursement of the revenue through lump-sum transfers assures us that the only distortions in the efficiency generated by the simulated policy are attributable to the environmental tax.

Results

In Figure 6.3, we present the effects of the different simulated policies on the Gross National Product (GNP), employment and social welfare. When the short-term environmental objectives pursued by the government are modest, with reductions in CO_2 emissions of less than 6 percent, small drops in the GNP of less than 0.5 percent will be produced. If we increase the reduction of emissions by an amount around 10–12 percent, the effects on GNP are significant, with drops of approximately 1 percent. Finally, the environmental objective proposed by the government for the period 2008–2012, a reduction in emissions of 16 percent, would have significant short-term effects on the economic activity equivalent to a loss of 1.6 percent of the GNP.

As such, there is a slight convex relationship between the reduction of emissions and the costs in terms of GNP. Small reductions in emissions are easily reachable, but these become costlier as the objectives of the environmental policy are toughened. These results highlight the opportunity costs of delaying the introduction of instruments for the control of Spanish CO_2 emissions. The less time it takes to reach a certain environmental objective, the higher the cost to the economy is. These results are consistent with the empirical evidence analyzed by different works, such as those by Grubb *et al.* (1993), Clarke *et al.* (1996) and Dean and Hoeller (1992).

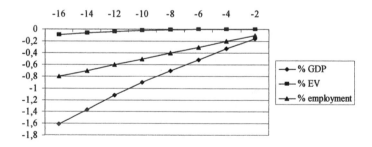

Source: own calculations.

Figure 6.3   Effects of percentage changes in CO_2 emissions with respect to 1995

The adverse effects of the environmental policy on economic activity translate into a drop in the demand for labour, as shown in Figure 6.3. The effects on employment, estimated through the model, were the results of two opposing economic forces. Firstly, the environmental objectives entail an additional cost for the different sectors, which are subject to the tax rate coming from the environmental regulation. Such extraordinary cost would have a negative effect on activity and, therefore, on the demand for labour. Secondly, the environmental regulation encourages labour demand. At the firm level, there is a substitution effect between dirtier and cleaner energies, but also between energy and other productive factors such as labour. At the macroeconomic level, the substitution effects can bring about sectorial changes in the structure of the economy, reducing the weight of energy-intensive sectors in favor of labour-intensive sectors.

The equilibrium between the previous opposite forces brings about a smaller reduction in the demand for labour when compared to economic activity. On the one hand, a drop in CO_2 emissions of less than 6 percent produces a negligible loss of employment that is below 0.3 percent. On the other hand, reducing Spanish CO_2 emissions by 16 percent would decrease employment by 0.8 percent, approximately half the effect caused on the GNP. These results contrast with those obtained by Manresa and Sancho (2005) and Faehn *et al.* (2005). The first paper estimates that a reduction of approximately 3.5 percent in Spanish CO_2 emissions would cause a 17 percent increase in unemployment, whereas the second indicates that a 25 percent cut in Spanish CO_2 emissions would have no effect on unemployment.

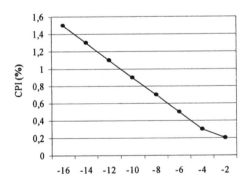

Source: own calculations.

Figure 6.4   Percentage changes in the Consumer Price Index (CPI)

Surprisingly, the effects of the simulated environmental policies on the Consumer Price Index (CPI) are similar to the effects on GNP, although they logically present the opposite sign. In Figure 6.4, we present relative changes with respect to the numeraire (international prices in our model). Small reductions in pollution emissions (lower than 6 percent) have very limited effects on inflation (less than 0.5 percent). In extreme cases, a 16 percent reduction in emissions would cause an increase of 1.5 percent in consumer prices.

Unlike with the variables previously studied, there is a strong non-linear relationship between the reduction in emissions and its effects on social welfare (excluding environmental effects). The changes in social welfare are measured in terms of Hicksian equivalent variations (EV), as is usual in the literature (Shoven and Whalley, 1992). Said relationship is convex; that is to say, the negative effects on social welfare grow more than proportionally than the changes in CO$_2$ emissions. In any case, the effects are of little significance. For instance, a reduction in emissions under 6 percent will cost less than 0.003 percent in EV, whereas a 16 percent reduction in CO$_2$ will cost only 0.1 percent in terms of social welfare. We should not forget that the fiscal revenues obtained by the environmental tax are returned to citizens through lump-sum transfers that reduce the impact of the public policy on consumption. Similar results have also been obtained by Böhringer *et al.* (1997) for a reduction of 20 percent in Spanish CO$_2$ emissions.

The costs for Spain of fulfilling the Kyoto terms could be smaller when environmental policies are articulated through green tax reforms. Using the revenues generated by environmental taxes to reduce other distortionary taxes generates a weak double dividend (Bovenberg and Goulder, 2002), which permits the reduction of the costs of environmental regulation. This is shown by diverse empirical works, such as Carraro *et al.* (1996), Capros *et al.* (1996) and Faehn *et al.* (2005). All of these have analyzed the effects of green tax reforms in Spain featuring a reduction in labour taxes. In all cases, the effects on GNP and employment were of little significance. However, Barker and Köhler (1998) estimate that a reduction in Spanish CO$_2$ emissions through a green tax reform could increase employment by 1.2 percent. Contrary to the previous results, Barbiker *et al.* (2001) estimate that, in terms of social welfare, the costs of Kyoto for Spain will be significant.

In order to study the sectorial effects of the environmental policy, we have analysed a policy of a 2 percent reduction in CO$_2$ emissions, a reasonable short-term objective.[8] From the simulation of said policy, we can observe the effects on the level of activity and the emissions in different sectors of our model, which are shown in Figure 6.5.

The most significant effects on production arise in primary energy sectors, with drops that oscillate between 2.8 percent for the coal sector and 2.1 percent for the refined petroleum sector. The electricity sector, however, experiences a slight drop equal to 0.4 percent, owing to two fundamental reasons. The classic thermal power utilities (coal, fuel oil, gas) represent only 40 percent of total electricity generation in Spain. In other words, only 40 percent of production from the electric sector will be indirectly levied by the environmental tax. In addition, this now makes electricity cheaper in relative terms with respect to fossil fuels, and that encourages electricity consumption through the substitution effects. The remaining non-energy sectors experience insignificant effects on their activity, ranging from a drop of 0.3 percent in diverse products to a null effect on agriculture and fishing, food, hotel and catering businesses and certain services (education, health, leisure and culture, etc.). The previous results are reasonable if we bear in mind that electricity represents approximately 70 percent of the final consumption of energy in Spain.

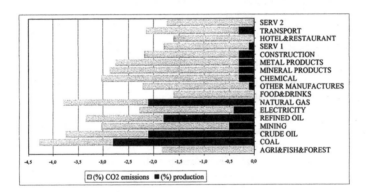

Source: own calculations[8].

Figure 6.5   Sectorial effects of a 2 percent reduction in CO_2

Emissions reduction is distributed in a less heterogeneous way than the changes in the level of activity. In general, all the economic sectors reduce their emissions in a significant way and by an amount over 1 percent. As expected, the drops in the energy sectors, over 3 percent except in electricity (2.3 percent), stand out. Other sectors such as diverse products (chemical, mineral and metallic products) also show large reductions, close to 3 percent, in their emissions. They are almost the same sectors regulated by the new European carbon market.

CONCLUSIONS

Spanish emissions of greenhouse gases have increased significantly over the last few years. This behaviour is incompatible with the objectives set up by the Kyoto Protocol and, in addition, it reflects an inefficient and very dependent energy system. In accordance with the internal distribution of the EU to comply with the Kyoto Protocol, Spain should reduce its greenhouse gas emissions by 16 percent if the calculations of the national allocation plan are reliable (however, at the moment of writing Spanish emissions continue to grow with respect to 1990 levels).

The objective of this chapter is to analyze the costs Spain would incur with these reductions through the introduction of a tax on CO_2 emissions. The methodology employed is a static AGE model for a small open economy. The consumption of energy goods on the part of industries and institutions has been broken down as much as possible from the national accounting data supplied. This characteristic gives the model sufficient flexibility so that the agents may substitute, in an efficient manner, the consumption of some energy goods for others that are less polluting. In addition, the model only simulates the CO_2 emissions associated with the consumption of fossil fuels. Therefore, the model produces cost-effective reductions of emissions.

The results reported in this chapter show that the costs of reaching the goals set by the EU for Spain are of little significance if the measures of environmental policy are taken far enough in advance. The objective of the government to reduce greenhouse gas emissions at an annual rate of 2 percent would result in a drop in the GNP equal to 0.2 percent, with this figure being even lower in terms of employment. In this case, the effects on social welfare would be null.

The energy sectors are logically the most affected by environmental regulation, with drops in activity of close to 2 percent. However, the consequences for the electricity sector are very small, owing to the effects of substitution which favour the consumption of this energy. The losses in activity in the remaining non-energy sectors are of little significance.

Nevertheless, should the Spanish government persist with its inhibited 'wait-and-see' policy, the economic consequences could be very negative. Given the limited temporary margin left by previous governments, an extreme situation could arise from a 16 percent reduction in emissions in the best-case short-term scenario (if the emissions maintain their current path, something improbable). In this context, the economic and social cost would be significant, with a loss of over 1 percent in GNP.

In sum, the conclusions are clear. It is especially advisable to introduce public measures for the control of greenhouse gas emissions as soon as possible. Obviously, these should be introduced through cost-effective instruments of environmental policy. For example, a hybrid system of taxes on some sectors and institutions along with an emissions trading system for other industrial sectors could be a good solution.

NOTES

* We have benefited from comments by Melchor Fernández, Alberto Gago, José M. Labeaga, Clemente Polo and participants in the First Atlantic Workshop on Energy and Environmental Economics. However, authors bear all responsibilities for any error or omission. Financial suport from the Spanish Ministry for Science and Technology and ERDF (Project SEC200203095), and the Galician government (Project PGIDIT03PXIC30008PN) is also acknowledged.

1. Some empirical evidence for the EU also include results for Spain, such as Carraro *et al.* (1996), Capros *et al.* (1996), Böhringer *et al.* (1997), Barbiker *et al.* (2001) and Barker and Köhler (1998).

2. See Labandeira and Rodríguez (2004) for a full description of the model.

3. It is also similar to GTAP-E (Rutherford and Paltsev, 2000) and MGS-6 for Norway (Faehn and Holmoy, 2003).

4. Based on an NAM published by Fernández and Manrique (2004) and national accounts (INE, 2002b).

5. The AGE model has been programmed in GAMS/MPSGE and we calibrated the model following the procedure in Rutherford (1999) by using the solver-algorithm PATH.

6. For a brief introduction to this methodology, see Shoven and Whalley (1992).

7. The wage elasticity of the labour supply was calibrated to be equal to -0.4, similar to that estimated for Spain by Labeaga and Sanz (2001). In order to gauge the elasticity of labour supply, we have followed the procedure used in Ballard *et al.* (1985), assuming, as in Parry *et al.* (1999), that leisure represents a third of the working hours effectively carried out in an initial equilibrium situation. We carried out a sensitivity analysis, increasing and reducing the labour elasticity by 50 percent, concluding that results from the AGE are robust.

8. MINING (Extraction of metallic, non-metallic nor energetic minerals); MINERAL PRODUCTS (Manufacturing of other non-metallic minerals, recycling); SERV1 (Telecommunications, financial services, real estate, rent, computing, R+D, professional services, business associations); SERV2 (Education, health, veterinary and social services, santation, leisure, culture, sports, public administrations).

REFERENCES

Ballard, C., J. Shoven and J. Whalley (1985), 'General equilibrium computations of the marginal welfare costs of taxes in the United States', *American Economic Review*, **75** (1), 128-138.

Barbiker, M., L. Viguier, J. Reilly, A.D. Ellerman and P. Criqui (2001), 'The welfare costs of hybrid carbon policies in the European Union', *MIT Joint Program on the Science and Policy of Global Change*, report 74.

Barker, T. and J. Köhler (1998), 'Equity and ecotax reform in the EU: Achieving a 10 percent reduction in CO₂ emissions using excise duties', *Fiscal Studies*, **19** (4), 375-402.

Böhringer, C., M. Ferris and T. Rutherford (1997), 'Alternative CO2 abatement strategies for the European Union', in S. Proost and J. Brader (ed.), *Climate Change, Transport and Environmental Policy*, Cheltenham: Edward Edgar, 16-47.

Böhringer, C. and T. Rutherford (1997), 'Carbon taxes with exemptions in an open economy: a general equilibrium analysis of the German tax initiative', *Journal of Environmental Economics and Management*, **32**, 189-203.

Bovenberg, L. and L. Goulder (2002), 'Environmental taxation and regulation', in A. Auerbach and M. Feldstein (ed.), *Handbook of Public Economics*, Dordrecht: Elsevier Science, 1471-1545.

Capros, P., T. Georgakopoulos, S. Zografakis, S. Proost, D. van Regemorter, K. Conrad, T. Schmidt, Y. Smeers and E. Michiels (1996), 'Double dividend analysis: First results of a general equilibrium model (GEM-E3) linking the EU-12 countries', in C. Carraro and D. Siniscalco (ed.), *Environmental Fiscal Reform and Unemployment*, Dordrecht: Kluwer Academic Publishers, 193-227.

Carraro, C., M. Galeotti and M. Gallo (1996), 'Environmental taxation and unemployment: Some evidence on the double dividend hypothesis in Europe', *Journal of Public Economics*, **62**, 141-181.

Clarke, R., G. Boero and A. Winters (1996), 'Controlling greenhouse gases: A survey of global macroeconomic studies', *Bulletin of Economic Research*, **48** (4), 269-308.

Dean, A. and P. Hoeller (1992), 'Costs of reducing CO₂ emissions: Evidence from six global models', *Working Paper 122*, Economics Department, OCDE.

Faehn, T., A. Gómez and S. Kverndokk (2005), 'Can a carbon permit system reduce Spanish unemployment?', *Discussion Paper 410*, Statistics Norway.

Faehn, T. and E. Holmoy (2003), 'Trade liberalisation and effects on pollutive emissions to air and deposits of solid waste. A general equilibrium assessment for Norway', *Economic Modelling*, **20**, 703-727.

Fernández, M. and C. Manrique (2004), 'La matriz de contabilidad nacional: un método alternativo de representación de las cuentas nacionales', *Documentos de Traballo da Area de Análise Economica 30*, IDEGA, Universidade Santiago Compostela.

Grubb, M., J. Edmonds, P. Brink and M. Morrison (1993), 'The costs of limiting fossil-fuel CO₂ emissions', *Annual Review Energy and Environment*, **18**, 397-478.

IEA (1998), *Energy Statistics of OECD Countries. 1995-1996*, Paris: International Energy Agency, OECD.

INE (2002a), *Estadísticas de Medio Ambiente. Cuentas Ambientales*, Madrid: Instituto Nacional de Estadística.

INE (2002b), *Contabilidad Nacional de España. Base 1995. Serie Contable 1995-2000. Marco Input-Output 1995-1996-1997*, Madrid: Instituto Nacional de Estadística.

Labandeira, X. and M. Rodríguez (2004), 'The effects of a sudden CO_2 reduction in Spain', *Documento de Traballo 0408*, Departmento de Economía Aplicada, Universidade de Vigo.

Labeaga, J. and J. Sanz (2001), 'Oferta de trabajo y fiscalidad en España. Hechos recientes y tendencias tras el nuevo IRPF', *Papeles de Economía Española*, **87**, 230-243.

Manresa, A. and F. Sancho (2005), 'Implementing a double dividend: Recycling ecotaxes towards lower labour taxes', *Energy Policy* **33** (12), 1577-1585.

Parry, I., R. Williams and L. Goulder (1999), 'When can carbon abatement policies increase welfare? The fundamental role of distorted factor markets', *Journal of Environmental Economics and Management*, **37**, 52-84.

Repetto, R. and D. Austin (1997), *The Costs of Climate Protection: a guide for the Perplexed*, New York: World Resource Institute.

Rutherford, T. (1999), 'Applied general equilibrium modeling with MPSGE as a GAMS subsystem: an overview of the modeling framework and syntax', *Computational Economics*, **14**, 1-46.

Rutherford, T. and S. Paltsev (2000), 'GTAP-Energy in GAMS: the dataset and static model', *Discussion Papers in Economics 00-02*, Center for Economic Analysis, University of Colorado at Boulder.

Shoven, J. and J. Whalley (1992), *Applying General Equilibrium*, Cambridge: Cambridge University Press.

7. An assessment of the consequences of the European emissions trading scheme for the promotion of renewable electricity in Spain

Pedro Linares, Francisco J. Santos and Mariano Ventosa

INTRODUCTION

In the last few years, the use of renewable energies in the European Union has increased considerably. The European Commission has certainly given this development political support by issuing the European Directive 2001/77/EC, which concerns the promotion of electricity produced from renewable energy sources, establishing indicative targets and some baseline regulation. However, not all countries are delivering at the same level. Basically, countries with a good rate of development are those using fixed tariffs or premiums, like Spain, Germany and Denmark. Evidence has shown that these types of instruments are much more effective for development than renewable portfolio systems (RPS) such as those implemented in the UK or Italy (Lauber, 2004). However, there is currently a discussion on whether this type of support is really cost-efficient compared with RPS. Although it seems that the price-based instruments may be more efficient dynamically (see, e.g., Menanteau *et al.*, 2003), there are still discussions on whether they are still too generous for renewable producers, and whether a renewable portfolio system would be more cost-effective in driving down costs for the consumer.

This situation may be even more acute with the recent approval of the European Directive 2003/87/EC, which establishes a scheme for greenhouse gas emission allowance trading within the Community (also known as European ETS). The ETS is expected to induce significant changes in

European electricity markets by imposing an implicit tax on carbon emissions from power plants. These changes include modifications in the operation of existing power plants, in electricity prices, in the revenues of existing firms (see for example Linares *et al.*, 2005), and in the profitability of new investments. Regarding this latter aspect, this implicit tax will indirectly encourage investments in 'clean' technologies such as gas, nuclear energy and renewables, while penalizing investments in other 'dirtier' technologies such as coal. This indirect incentive for renewables is of special importance in the context of the debate already mentioned, since it will add to the profits of renewable producers through a higher electricity price. Although it is debatable whether the premiums for renewables account for their carbon reduction benefits or for their local positive externalities (which most of the times is not so clear; see, e.g., Komor and Bazilian, 2005), it is possible that the implicit carbon tax may in fact be double-counting the carbon externality of renewables because of the slow adjustment of premiums (while green certificates are automatically adjusted). Therefore, current premium or fixed tariff levels might have to be revised under this new scenario.

The objective of this chapter is to analyze this impact quantitatively by simulating the operation and expansion of the Spanish electricity sector with and without the EU ETS, and specifically, to look at the combined effect of the ETS and different renewable support systems (premiums and RPS) in Spain, in order to determine whether or not the current support policies will have to be revised. Spain is considered a good reference for the analysis because of the large degree of renewables development, and also because of the large carbon reductions it needs to comply with the Kyoto Protocol.

Some analyses have already been carried out on the expected consequences of carbon trading mechanisms on renewable support schemes. For instance, Jensen and Skytte (2003) studied the interaction between carbon trading and tradeable green certificates using an analytical model. Del Río *et al.* (2005) have looked at the impact of clean development mechanisms and joint implementation on the deployment of renewable electricity in Europe, although again from an analytical point of view. Hindsberger *et al.* (2003) used a simulation model to obtain quantitative results for the Nordic electricity market also under a tradeable green certificate system. However, this model has some limitations when applied to real electricity markets.

In this chapter, estimates have been produced by using an oligopolistic generation expansion model for the Spanish electricity sector, the ESPAM model, which has been developed at our Institute. This model is described in section 2, and the results obtained are presented in section 3. Finally, section 4 provides the conclusions drawn from the study.

THE MODEL

The electricity market is modelled as one in which, in the short term, firms compete in quantity of output as in the Conjectural Variations (CV) approach in which generators are expected to change their conjectures about their competitors' strategic decisions, in terms of the possibility of future reactions (Ventosa *et al.*, 2005). In the long-term electricity market, firms compete in generating capacity as in the Nash game. Conceptually, the structure of this model corresponds to various simultaneous optimizations – for each firm, the maximization of its profits subject to its particular technical constraints. These optimization problems are linked together through the electricity price and the emissions permit price resulting from the interaction of all of them. The general structure of the model is shown in Figure 7.1.

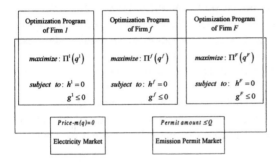

Source: authors' elaboration.

Figure 7.1   Mathematical structure of the market equilibrium model

The objective of each generation firm *f* is to maximize its profit Π defined as market revenues minus operating costs, investment costs and cost of purchasing emission permits. The sets of constraints *h* and *g* ensure that each company's optimization program provides decisions *q* that will be technically feasible. The link between the electricity market and every firm's optimization program is the demand function that relates the demand supplied by every generator to the electricity price. It is assumed in this model that the total demand at each load level is a linear function of the price.

The permit market is modelled as a perfectly competitive one. So the clearing price of the market will be the crossing of the permits' aggregated demand curve with the supply curve. The supply curve is set to be a constant quantity of permits determined by the government. The aggregated demand curve is the sum of demands of every company at every permit price (which is in turn based on its marginal abatement costs). The link between the permits market and every firm's optimization program is the constraint that sets the emission of pollutants to the amount of permits owned by each company. The dual variable of such constraint is the permit price.

The model assumes that firms make their capacity-expansion decisions as in a Nash equilibrium. Formally, the investment market equilibrium defines a set of capacities such that no firm, taking its competitors' capacities as given, wishes to change its own capacity unilaterally (Ventosa *et al.*, 2002). Thereby, each firm chooses its new maximum capacity so that its own profit is maximized. The Nash assumption implies that firms' investment decision-making occurs simultaneously.

To summarize, the whole model, which is a CV-market sub-model plus a Nash-expansion planning sub-model, subject to the environmental restriction of the permits market, defines the operation, the investment, permits purchases and pricing of both electricity and permits that simultaneously satisfy the first-order optimality conditions of all firms and that of the permits market.

This market equilibrium problem can be stated in terms of a Linear Complementarity Problem (Rivier *et al.*, 2001) by means of setting the first-order optimality Karush-Kuhn-Tucker conditions associated to the set of maximization programs. A more detailed description of the model can be found in Linares *et al.* (2005) and a full mathematical structure is presented in Lapiedra *et al.* (2003).

AN APPLICATION TO THE SPANISH ELECTRICITY SECTOR

The model described above has been used to simulate the impact of the European ETS on the new investments and profitability of renewable energies in the Spanish electricity sector (for a description of the Spanish electricity sector, see, e.g., Crampes and Fabra, 2005). First the general conditions for the simulation are shown, and then the specific results on investments are presened.

General conditions for the simulation

The case study analyzes the expansion of the Spanish electricity system for the next 16 years (2005–2020). The six existing generating firms have been considered (although their names have been omitted), plus other possible new entrants and the special regime (that is, renewable energy producers). The investment capacity of the firms has been limited to a different number of power plants to be built in a certain period of five years.

All the power plants belonging to the generators have been aggregated into one group per technology and firm, in order to reduce the size of the model. The existing technologies considered are nuclear (NCL), fuel (FO), natural gas (GN), gas combined cycles (ECCGT), domestic coal (HLL), imported coal (CI), brown lignites (LGP), black lignites (LGN), regulating hydro (REG), run-of-the-river hydro (FLU), pumping units (BOMB), biomass (EBIO), cogeneration (ECOG), small hydro (EMINH), wind (EEOL), and solar (ESOL). In addition, future technologies have been considered for the new investments: supercritical coal (CSC), advanced nuclear (NCLAV), gas combined cycles (CCGT), three types of biomass (BIO1: energy crops, BIO2: agricultural waste, and BIO3: forest waste), three types of wind depending on the wind speed of the site (EOL1, EOL2, EOL3), small hydro (MINH), cogeneration (COG), and solar thermal (SOLT). Their parameters are presented in Tables 7.1–7.4.

As for the carbon emissions market, the amount of permits allowed is that established by the Spanish National Allocation Plan (NAP) (RD 60/2005), that is, 160 Mt. However, as said before, the Plan only covers the period 2005–2007. From 2008, the government envisions that emissions should not be higher than those of 1990 incremented by 24 percent (a 15 percent increase over 1990, plus 7 percent obtained from clean development mechanisms, plus 2 percent from carbon sinks), so the total amount of permits from 2008 to 2014 should be 147.8 Mt.

It has to be noted that this is the whole amount of permits distributed among all sectors covered by the ETS Directive. However, only the electricity sector has been modelled in detail. The rest of the sectors are much more difficult to model adequately because of their disaggregation (there are many small CO_2-producing facilities, with very different characteristics) and lack of data. However, this same disaggregation allows us to assume that they will behave as price-takers in the emissions market, and therefore they may be modelled as a competitive fringe by means of a residual demand function. This demand function is the aggregated marginal abatement cost curve for all these sectors in Spain, and has been obtained from the PRIMES model (Capros *et al.*, 2001).

Table 7.1   Parameters for current thermal power plants

FIRM	TECHNOLOGY	LINEAR VARIABLE COST	QUADRATIC VARIABLE COST	INSTALLED POWER	CO_2 EMISSIONS RATE
		c€/MWh	c€/MW²h	MW	t/MWh
1	NCL-1	330	0.02	3358	0
	HLL-1	1500	0.41	1021	0.95
	CI-1	1560	0	220	0.90
	FO-1	4140	0.13	2337	0.78
	GN-1	3930	0.35	830	0.79
	CCGT-1	2173	0.16	1500	0.40
2	NCL-2	330	0.02	3641	0
	HLL-2	1785	0.06	1462	0.96
	LGP-1	1845	0.07	1469	0.99
	LGN-1	1986	0	1100	0.93
	CI-2	1380	0.02	1712	0.92
	FO-2	4440	2.12	400	0.77
	GN-2	4065	0.16	1543	0.72
	CCGT-2	1978	0.14	1200	0.40
3	NCL-3	330	0.03	739	0
	HLL-3	1530	0.23	1498	0.90
	LGP-2	1725	0	583	1.27
	FO-3	4140	0.74	447	0.76
	GN-3	4218	0	155	0.99
4	HLL-4	1881	0.06	544	0.90
	LGN-2	1788	0.52	400	0.94
	FO-4	3690	0.35	682	0.76
5	NCL-4	348	0	165	0
	HLL-5	1470	0.41	1588	0.92
	CCGT-3	2554	0	450	0.40
6	CCGT-4	1978	0.20	800	0.40
OTHER	CCGT-5	2304	0	400	0.40

Source: own elaboration based on CSEN (1997).

Of course, since the ETS Directive sets up a European permit market, allowing trade between countries, the expected results for the allowance price may be different than those simulated here. On the one hand, the Spanish energy technologies and the energy mix are similar to the European average, so the marginal abatement costs curves (which ultimately define the price of the allowance) would be expected to be similar. However, since the size of the market will be much larger when enlarged to a European scale, abatement opportunities may increase and therefore the expected price of the permit should be lower. The problem is that this European market has not been defined yet, since there are still some countries without a definitive NAP (and therefore, the total amount of permits to be distributed in Europe is unknown).[1] Therefore, we considered it more advisable to simulate just the Spanish emissions market with no trade, assuming that the real permit price may be somewhat lower. This of course will influence renewables development, which will be lower with a lower permit price.

Table 7.2   Parameters for renewables and cogeneration power plants

FIRM	TECHNOLOGY	LINEAR VARIABLE COST	INSTALLED POWER	USE RATIO	PREMIUM	CO_2 EMISSIONS RATE
		c€/MWh	MW		c€/MWh	t/MWh
Special regime	EBIO	781	436	0.413	2919	0
	ECOG	2887	5785	0.319	2128	0.55
	EMINH	0	1637	0.305	2946	0
	EEOL	0	7782	0.211	2664	0
	ESOL	0	16	0.107	12020	0

Source: own elaboration based on CSEN (1997).
Use ratio: since renewable power plants are not able to produce continuously, the use ratio expresses the relationship between the installed power and the energy produced.
Premium: the amount paid by the government to renewables and cogeneration over the market price.

Table 7.3   Parameters for current hydro power plants

FIRM	REG		FLU		BOMB	
	MAXIMUM POWER (MW)	ANNUAL INFLOWS (GWh)	MAXIMUM POWER (MW)	MAXIMUM POWER (MW)	PUMPING YIELD (%)	MAXIMUM CAPACITY (GWh)
1	3150	8930	360	628	70	300
2	2100	2839	390	1409	70	515
3	850	1538	188	208	70	90
4	475	243	41	340	70	50
5	270	264	38	0	70	0
6	0	0	0	0	70	0
OTHER	0	0	0	0	70	0

Source: own elaboration based on CSEN (1997).

Table 7.4   Parameters for new technologies

TECHNOLOGY	LINEAR VARIABLE COST	INVESTMENT COST	MAXIMUM POWER	USE RATIO	PREMIUM	CO_2 EMISSIONS RATE
	c€/MWh	€/KW	MW		c€/MWh	t/MWh
CCGT	2100	466		1		0.40
NCLAV	790	2000		1		0
CSC	1500	992		1		0.80
BIO1	5017	1272	1131	0.799	2932	0
BIO2	1003	1406	1212	0.799	2932	0
BIO3	6688	1142	687	0.799	2932	0
MINH	0	2700	743	0.267	2932	0
COG	4700	600	1315	0.426	2199	0.63
EOL1	0	900	2444	0.247	2932	0
EOL2	0	900	3665	0.212	2932	0
EOL3	0	900	6109	0.159	2932	0
SOLT	0	6000	200	0.109	18326	0

Source: own elaboration based on European Commission (2004).

As mentioned before, the objective of this chapter is to analyze the impact of the EU ETS on the development of renewables. To that end, the model has been used to simulate the expansion of the Spanish electricity system with and without the ETS, considering that the current support scheme for renewables (based on premiums for electricity produced) is maintained. We have also analyzed the impact of a substitution of premiums by a tradeable green certificate system in order to analyze whether the premiums may be too generous, and we have also looked at the higher renewable penetration scenario envisioned by the European Directive 2001/77 on renewables. Therefore, the three cases considered (with and without the Carbon Directive) are:

- Renewable promotion through premiums, as in the current scheme.
- A renewable quota equivalent to the one that would be reached under the premium mechanism. This is done in order to compare the impacts of the two mechanisms on the electricity market, provided that the same amount of renewables is achieved.
- A renewable quota so that in 2010 17.5 percent is reached, according to the European Directive 2001/77 on renewables.

As may be expected, results are very sensitive to the participation of nuclear energy in the system. The assumption has been made that, for the business-as-usual case, investments in nuclear energy are not attempted because of the current investment risks linked to this technology. But a sensitivity analysis considering the participation of nuclear energy has been carried out.

Regarding other relevant assumptions for the model, the annual growth of electricity demand has been set as an annual average of 2.5 percent, based on the estimations of the Spanish government (MINER, 2002). The slope of the electricity demand curve has been set at 600€/MWh.MW. Also, a residual demand curve slope of 1.3€/MWh.MW has been considered for the two largest firms. The discount rate used for investments is 9 percent.

Results

In this section, the major indicators regarding renewables penetration are analyzed for all cases studied and presented in Table 7.5 and 7.6. These results have been obtained by running the model presented before. The model has been programmed in GAMS language. The mixed complementarity problem has been solved with the PATH solver, and the cost minimisation problem with the CPLEX solver.

It has to be remarked that all monetary values shown correspond to the 2004 euro, and that no inflation has been considered in order to better observe the impact of the instruments analyzed.

Table 7.5   Installed power in 2020 per technology (MW)

	WITHOUT ETS DIRECTIVE		WITH ETS DIRECTIVE	
	PREMIUMS	RPS 17.5%	PREMIUMS	RPS 17.5%
CCGT	9988	7310	16256	12723
NCLAV	-	-	-	-
CSC	213	-	-	-
BIO1	-	1021	-	1021
BIO2	1094	1094	1094	1094
BIO3	-	225	-	225
COG	-	-	-	-
EOL1	2206	2206	2206	2206
EOL2	3308	3308	3308	3308
EOL3	-	5513	2315	5513
TOTAL	16808	20676	25179	26089

Source: own calculations.

We see that, in the base case (premiums, no ETS Directive), there is a significant development of renewable energies, basically wind energy at good and average sites, and agricultural residues. When the ETS Directive is introduced, more wind energy is installed (at less-than-average sites), but no more renewables are promoted. In fact, most of the impact of the Directive on new investments is on gas combined cycles, rather than on renewables. The impact of the Directive on renewable investments is even decreased when the quota of renewables is higher: we see that no new renewable investments are indirectly promoted by the ETS Directive if a 17.5 percent quota is set up for 2010. The situation is quite similar if we look at the energy produced in Table 7.6.

As seen previously for installed power, the share of electricity produced during the simulation period by renewables is not affected very much by the introduction of the ETS Directive, only a small participation of wind is promoted, and this only if the amount of renewables does not reach the objectives proposed by the European Commission. Again, the major impact of the Directive is not on renewables, but on the substitution of domestic coals by gas combined cycles.

However, an interesting effect may be seen in this table, in that when the renewables quota is increased, and there are carbon reduction policies, there is less substitution of coal by gas. Since the increase of renewables already provides for some of the carbon reductions needed, the price of the carbon permit decreases and therefore there is less advantage for changing to gas. So indirectly, a stronger promotion of renewables is helping domestic coal to compete against gas.

Table 7.6   Electricity produced in 2005–2020 per technology (percent total energy)

	WITHOUT ETS DIRECTIVE		WITH ETS DIRECTIVE	
	PREMIUMS	RPS 17.5%	PREMIUMS	RPS 17.5%
NCL	24.48%	24.48%	24.48%	24.48%
HLL	16.13%	16.42%	10.18%	11.15%
LGP	5.85%	5.98%	2.26%	2.41%
LGN	3.88%	4.03%	1.06%	1.21%
CI	5.98%	5.98%	5.47%	5.88%
FO	0.01%	0.00%	0.00%	0.00%
ECCGT	4.67%	4.70%	6.14%	6.28%
EBIO	0.56%	0.56%	0.56%	0.56%
ECOG	5.72%	5.72%	5.72%	5.72%
EMINH	1.55%	1.55%	1.55%	1.55%
EEOL	5.09%	5.09%	5.09%	5.09%
ESOL	0.01%	0.01%	0.01%	0.01%
REG	5.41%	5.41%	5.41%	5.41%
BOMB	0.00%	0.00%	0.00%	0.00%
FLU	3.49%	3.49%	3.49%	3.49%
CCGT	9.32%	6.72%	20.29%	16.89%
NCLAV	0.00%	0.00%	0.00%	0.00%
CSC	0.59%	0.00%	0.00%	0.00%
BIO1	0.00%	0.81%	0.00%	0.81%
BIO2	3.00%	3.00%	3.00%	3.00%
BIO3	0.00%	0.05%	0.00%	0.05%
COG	0.00%	0.00%	0.00%	0.00%
EOL1	1.87%	1.87%	1.87%	1.87%
EOL2	2.41%	2.15%	2.41%	2.15%
EOL3	0.00%	2.00%	1.03%	2.00%
TOTAL	100.0%	100.0%	100.0%	100.0%

Source: own calculations.

As mentioned previously, some of these impacts may be explained by looking at the prices of the electricity market with and without the Directive, as shown in Table 7.7. We see that electricity prices increase 9 percent in 2012 and 27 percent in 2020 due to the introduction of the ETS Directive, thus making gas competitive, and increasing the revenues of electricity producers. However, this increase in electricity prices is not enough to compensate other renewable technologies such as energy crops, forest residues, or solar thermal, which have a very large potential.

It may also be observed that the price increase is smaller when the renewables quota increases (due to the marginal price system; see, e.g., Jensen and Skytte, 2003). This explains why domestic coal is not often substituted by gas in this case.

So we see that electricity prices increase, but do not promote a significant growth of renewables. However, they do increase the revenues of the installed facilities (some of which are profitable renewables such as good wind sites or agricultural residues). This may be seen in Tables 7.8 and 7.9 by looking at the equivalent price of green certificates, and by looking at the revenues of renewable producers.

Table 7.7   Electricity market price (€/MWh)

	WITHOUT ETS DIRECTIVE		WITH ETS DIRECTIVE	
	PREMIUMS	RPS 17.5%	PREMIUMS	RPS 17.5%
2005	24.90	25.03	24.90	25.03
2006	25.24	25.30	25.24	25.30
2007	25.79	25.66	25.79	25.66
2008	26.31	26.11	28.65	28.78
2009	26.96	26.68	28.85	28.92
2010	26.98	27.01	29.02	29.07
2011	27.01	27.05	29.18	29.23
2012	27.04	27.08	29.36	29.41
2013	27.10	27.08	30.10	29.69
2014	27.11	27.09	30.37	29.91
2015	27.11	27.14	30.65	30.15
2016	27.16	27.19	30.98	30.42
2017	27.23	27.26	31.33	30.71
2018	27.29	27.28	33.38	32.11
2019	27.30	27.28	33.92	32.56
2020	27.30	27.28	34.56	33.05

Source: own calculations.

Table 7.8   Green certificate prices under the RPS 17.5 percent scenario (€/MWh)

	WITHOUT ETS DIRECTIVE		WITH ETS DIRECTIVE	
	RPS AS PREMIUMS	RPS 17.5%	RPS AS PREMIUMS	RPS 17.5%
2005	28.01	14.98	15.21	14.98
2006	27.78	14.72	27.78	14.72
2007	27.32	14.36	27.86	14.36
2008	26.91	27.25	25.05	24.57
2009	26.07	26.68	23.86	24.43
2010	25.79	26.35	24.25	24.28
2011	25.74	26.31	24.09	24.12
2012	**25.73**	**26.27**	**23.90**	**23.94**
2013	25.72	40.73	40.25	38.12
2014	25.67	40.73	34.58	37.90
2015	25.60	40.68	33.91	37.66
2016	25.53	40.63	21.78	37.40
2017	25.53	40.56	50.20	37.11
2018	25.53	40.54	19.58	35.71
2019	25.53	56.59	19.01	51.31
2020	**25.52**	**56.59**	**33.42**	**50.82**

Source: own calculations.

First we take a look at green certificate prices. Recall that these prices are the differential between the long-run marginal cost of renewables and the electricity market price. We have analysed this price for a situation equivalent to the current one (with premiums), by setting a quota equal to the one obtained with premiums, and also for the case in which the renewable quota set by the European Commission is enforced. We can only compare green certificate prices for the RPS 17.5 percent case, since in the RPS case equivalent to premiums the share of renewables is different, and therefore it will be impossible to isolate the impact of the ETS, as the green certificate price will incorporate both the impact of the ETS Directive and the different amount of energy produced by renewables.

We see that the introduction of the ETS Directive in the RPS 17.5 percent case decreases green certificate prices. That is, the additional cost to be paid to renewable producers is lower (since the electricity price is higher, renewable producers are able to recover a larger part of their costs in the electricity market). Therefore, if premiums are not adjusted before the introduction of the ETS Directive, there will possibly be an overexpenditure on renewables. This is even more evident if we realise that green certificate prices under the current situation are lower than the current premiums, so we might think that there is already an overexpenditure (see a comment on this in the section discussing revenues).

Table 7.9   Revenues of renewable producers per technology (M€, net present value 2005–2020)

	WITHOUT ETS DIRECTIVE			WITH ETS DIRECTIVE		
	PREMIUMS	RPS AS PREMIUMS	RPS 17.5%	PREMIUMS	RPS AS PREMIUMS	RPS 17.5%
EBIO	707	681	711	738	714	711
EMINH	1971	1889	1972	2056	1980	1972
EEOL	6150	6212	6486	6429	6513	6486
ESOL	18	6	7	18	7	7
BIO1	0	0	819	0	0	819
BIO2	3813	3663	3825	3977	3841	3825
BIO3	0	0	43	0	0	43
EOL1	2377	2284	2384	2479	2394	2384
EOL2	3060	2939	2673	3191	3082	2673
EOL3	0	0	2309	1198	1189	2309
TOTAL	18097	17674	21229	20086	19721	21229

Source: own calculations.

We also see that, when a higher quota is enforced, green certificate prices are higher, but also decrease because of the introduction of the ETS Directive. But in all cases, they are higher than the current premiums paid for renewables (from 2012). So another conclusion is that, in order to achieve the 17.5 percent objective, current premiums may not be enough as of 2012.

As seen in Table 7.6, the same amount of energy is produced by all technologies under the current support scheme (premiums) with and without the ETS Directive, except for EOL3. However, these technologies receive more revenues because of the increase in electricity prices (18,888M€ compared to 18,097M€). Therefore, there may be a certain inefficiency of the support system, given that premiums are determined based on a desired share of renewables. If the same share of renewables is to be maintained under the ETS Directive, that should not cost more for the consumer. Therefore, premiums should have to be revised.

An RPS system, on the contrary, adjusts automatically to electricity price increases. We see that clearly in the RPS 17.5 percent case, which produces the same renewable energy under both scenarios, and in which we see that the revenue is exactly the same: the RPS system adjusts automatically by reducing green certificate prices, thus keeping revenues at the same level (the one required to meet the quota).

But of course, in these cases, we are assuming that the share of renewables is maintained. If it is to be increased, then costs may also have to increase, although care should be taken so that they are not excessive. This is shown in the RPS case equivalent to premiums which presents two effects: on the one hand, since the amount of renewables is higher, the marginal cost of these technologies is also higher, and this increases revenues to all producers (and therefore costs to consumers), given that this marginal cost determines the price paid to all of them under an RPS system. But on the other hand, there is also an automatic adjustment in the price paid to reflect the increase in electricity prices. In the end, the cost for the consumer (which equals the revenue of renewable producers) is higher under the premium scheme than under the equivalent RPS scheme, both with and without the ETS, which again points to a certain inefficiency of premiums or, at least, a certain inflexibility of this instrument.

Of course, it must be remembered that this inefficiency is measured in static terms, whereas in a dynamic context premiums may prove superior to RPS, as argued by Menanteau *et al.* (2003). Also, RPS have larger risks, which have not been modelled here, and which should also be considered when deciding between instruments. However, our main argument here is not that premiums are inferior to RPS, but only that premiums should be adjusted due to the introduction of the ETS Directive in order not to increase consumer costs excessively if the renewables share is to be maintained.

Finally, we have looked at the impact of the Directive when new nuclear investments are undertaken. When nuclear energy is considered an option for new investments, it does not hold a very large share in the system, just 2.3 percent of energy and some 800 MW installed. It substitutes the less profitable wind promoted by the ETS Directive, and also some gas. But basically, results are not affected much; nuclear energy is not able to compete with renewables if renewable support systems are maintained.

The sensitivity of the results to other parameters such as gas prices or the amount of carbon permits distributed is not significant, as shown in other previous analyses (Linares *et al.*, 2005).

CONCLUSIONS

This chapter has presented some estimations of the impact of the European ETS Directive on the promotion of renewable energies in the Spanish electricity sector by using a detailed oligopolistic model for simulating the expansion and operation of its generating facilities.

Results show, that, contrary to the expectations of some sectors, the Directive will not induce significant changes in the profitability of renewable energies and therefore renewable installed power will not grow very much (except for some close-to-profitability wind sites). In fact, what the Directive does indirectly stimulate is new investments in gas combined cycles. This may be explained by the evolution of the electricity prices which incorporate the implicit carbon tax set up by the carbon trading regime. The increase of price is significant, but not enough to promote additional investment in new renewable technologies. Therefore, the emissions trading scheme will not be able to substitute current renewable support measures, which should be maintained.

However, although renewables are not developed much further, the Directive does increase revenues for producers, because of the increase in electricity prices. For most renewable energy technologies, the same energy is produced, but a higher price is paid for it, so the efficiency of the system may be questioned (although it may be argued that this higher price may serve as a long-term signal for new investments, or as a risk premium). In fact, if we look at the evolution of green certificate prices, we see that they are reduced by the introduction of the ETS Directive. Therefore, it seems that premiums should be revised in order not to provide excessive revenues to producers if the renewable share is to be maintained. On the other hand, we also see that, because of this low impact of the ETS Directive, the 17.5 percent renewable target established in the European Directive for renewables cannot be met with price-based instruments unless premiums are increased significantly.

Therefore, if current renewable penetration is to be maintained, premiums should be reduced in order not to be an excessive cost for consumers due to the introduction of the ETS Directive. But if a more ambitious target is pursued, they should rather be increased, although always taking into account that, under the ETS Directive, conditions will be somewhat more favourable and therefore premiums lower than without the Directive. Here RPS present some advantages compared to premiums because of their automatic adjustment to changes in electricity market prices.

To summarize, the ETS Directive will not significantly alter the development of renewable energies, but may cause annoying effects on the efficiency of the support systems already in place when these are price-based. Regulators will have to examine this issue thoroughly as the carbon emissions market develops.

NOTES

* This chapter has been supported in part by Fundación BBVA and by the European Commission (Contract 4.1030/C/02 004/2002).

1. It must be remembered that the EU ETS Directive does not cover all carbon emissions, and therefore does not equal the reductions imposed by the Kyoto Protocol.

REFERENCES

Capros, P., N. Kouvaritakis and L. Mantzos (2001), 'Economic Evaluation of Sectoral Emission Reduction Objectives for Climate Change. Top-down analysis of green house gas emission reduction possibilities in the European Union', National Technical University of Athens.

Crampes, C. and N. Fabra (2005), 'The Spanish Electricity Industry: Plus ça change ...', *The Energy Journal*, Special Edition European Electricity Liberalisation, 127-153.

CSEN (1997), 'Una simulación del funcionamiento del Pool de Energía Eléctrica en España', Dirección de Regulación, Comisión del Sistema Eléctrico Nacional.

Del Río, P., F. Hernández and M. Gual (2005), 'The implications of the Kyoto project mechanisms for the deployment of renewable electricity in Europe', *Energy Policy*, **33**, 2010-2022.

European Commission (2004), 'Sustainable Energy Technology Reference Information System (SETRIS)', Joint Research Centre, European Commission, http://www.jrc.es.

Hindsberger, M., M.H. Nybroe, H.F. Ravn and R. Schmidt (2003), 'Co-existence of electricity, TEP and TGC markets in the Baltic Sea Region', *Energy Policy*, **31**, 85-96.

Jensen, S.G. and K. Skytte (2003), 'Simultaneous attainment of energy goals by means of green certificates and emissions permits', *Energy Policy*, **31**, 63-71.

Komor, P. and M. Bazilian (2005), 'Renewable energy policy goals, programs, and technologies', *Energy Policy*, **33**, 1873-1881.

Lapiedra, L., M. Ventosa and P. Linares (2003), 'Expansion planning model considering an emission-based permits market', Proceedings 8th Portuguese-Spanish Congress on Electrical Engineering, Vol 2. Vilamoura (Portugal), 3-5 July.

Lauber, V. (2004), 'REFIT and RPS: options for a harmonised Community frame work', *Energy Policy*, **32**, 1405-1414.

Linares, P., F.J. Santos, M. Ventosa and L. Lapiedra (2005), 'Estimated impacts of the the Spanish electricity sector', *The Energy Journal*, **27**, 79-98.

Menanteau, P., D. Finon and M.L. Lamy (2003), 'Prices versus quantities: choosing policies for promoting the development of renewable energy', Energy Policy, **31**, 799-812.

MINER (2002), 'Planificación de los sectores de electricidad y gas', Ministerio de Economía y Hacienda.

Rivier, M., M. Ventosa and A. Ramos (2001), 'A Generation Operation Planning Model in Deregulated Electricity Markets based on the Complementarity Problem', in M. Ferris, O. Mangasarian, and J. Pang (eds), Applications and algorithms of complementarity, Kluwer Academic Publishers.

Ventosa M., A. Baillo, A. Ramos and M. Rivier (2005), 'Electricity Market Modelling Trends', *Energy Policy*, **33**, 897-213.

Ventosa, M., R. Denis and C. Redondo (2002), 'Expansion Planning in Electricity Markets. Two Different Approaches', Proceedings 14th PSCC Conference, Seville, July.

8. Efficient verification of firm data under the EU emissions trading system

Frauke Eckermann

INTRODUCTION

In January 2005, the EU-wide greenhouse gas (GHG) emissions trading scheme (EU, 2001) officially started operating. Under this trading scheme, the amount of allowances that a firm initially receives is dependent on the firm's own report of its actual emissions. High verification standards are essential to prevent firms from having an advantage from overreporting their actual emissions. On the other hand, too-strict verification rules lead to high costs.

The trading scheme itself is divided into several temporal stages, beginning with a three year trading period from 2005 to 2007. The second trading period lasts from 2008 to 2012, and coincides with the commitment period under the Kyoto Protocol (UNFCCC, 1997). This is followed by further five year trading periods. The launch of the trading scheme before the legally binding phase under the Kyoto Protocol will enable European companies (and countries) to get acquainted with the instrument and, if necessary, allows the European Commission to make adjustments in time. The 'currency' of the scheme is the so called GHG allowances, measured in tons of CO_2 equivalents.[1] In the first period of the EU emissions trading scheme, only CO_2 will be traded. The system will be enlarged later to include the other GHGs. Trading will take place among companies. This is different from the trading scheme under the Kyoto Protocol, where trading is foreseen only at the state level. Participation in the trading programme is mandatory for large installations of energy-intensive industries. The total quantity of allowances is limited, aiming to fulfil the Kyoto targets, which constitute a reduction of 8 percent of 1990 emissions over the period from 2008 to 2012 for the European Union.

At the beginning of each trading period, a certain amount of allowances will be allocated to each firm free of charge. This amount is dependent on its own report of its current emissions. Since these reports need not necessarily present the true data, they need to be verified. With respect to the verification of firm data, the EU Directive (EU, 2004) states that strict verification standards are essential and that penalties for infringements shall be assigned by the Member States.

In view of the tight schedule for the allocation of allowances and of the verification costs, however, it seems unlikely that all plants can be audited sufficiently and thoroughly.

Even though there is a lot of literature on the comparison of permit trading schemes with other policy instruments,[2] as well as emissions trading between states[3] and among companies[4] in the context of the Kyoto Protocol, there is, to my knowledge, not much literature on verification problem yet. The literature on the verification of firm data concentrates mainly on technical or legal issues, like the study by the Center for Clean Air Policy (2001), but not on economic efficiency. The verification of firm data seems to be in the field of compliance and enforcement problems,[5] and in particular related to models with self-reporting on the compliance status, where the verification is conditioned on these reports (Kaplow and Shavell, 1994). There is, however, an important difference to the verification problem which is analyzed in this chapter. I do not consider compliance problems, but problems related to the initial allocation of allowances. Firms incur no costs for reporting false data, unless they are audited and penalized.

The scenario in the present article is such that firms receive subsidies in the form of GHG allowances which are based on their emissions reports. Firms hold private information on their emissions data. In addition, it seems reasonable that they do not take their rivals' actions into consideration. Therefore, the problem resembles the one of tax evasion, where firms have to pay a tax which is based on their (voluntarily) reported income. Following the approach that Chander and Wilde (1998) have used for the characterization of optimal income tax and enforcement schemes, I apply a verification method consisting of three elements in this analysis. Firstly, allowances are allocated on the basis of the reports of the firms *via* an allocation function. Secondly, audit probabilities, based as well on the reported data, are fixed for each firm. Thirdly, penalties are imposed for misreporting. In the model, auditing is costly and firms and the government are risk-neutral. Penalties are exogenously constrained. I analyze two different penalty structures. One applies a penalty that is linear in the degree of misreporting, the other applies a maximum penalty for each incorrect report.

Instead of specifying an objective function for the government, I introduce the notion of an efficient scheme. This captures the fact that the mechanism is likely to be set up by a consortium consisting of members of the government as well as industry. The latter will then wish to maximize the amount of overall allowances allocated, while the former wants to keep audit costs low. An efficient scheme will have the property that it is not possible to allocate the same amount of allowances with less auditing.

The proposed verification mechanism should be preferred to full verification, i.e., an audit probability of one for each firm, due to cost efficiency. Under a scheme with maximum penalties, full verification is never efficient. Under the linear penalty scheme, there is only one allocation function for which full verification is efficient. This is the function that provides firms with an amount of allowances equal to the amount of emissions they have reduced. Verification mechanisms with random auditing, on the other hand, lead to efficiency for a broader set of functions. This result parallels the result in the tax evasion context, namely, that optimal schemes typically involve random auditing (see Border and Sobell, 1987). I derive and discuss the interplay between the form of the allocation function and the audit probabilities. For concave allocation functions, which correspond to a regressive allocation of emission allowances, the audit probability of an efficient verification mechanism is nonincreasing in the reported values. It is dependent on the firm's average allocation of emission allowances. In case of a proportional allocation of allowances, which can be observed in most countries, the audit probability should be the same for all firms.

The chapter starts with the setup of the model and the introduction of the notion of efficiency. I characterize efficient verification schemes with linear penalties, and compare them to full verification before I extend the analysis to a setting with maximum penalties. Finally, I present concluding remarks.

THE MODEL

In this analysis, I study the problem that the initial amount of allowances a firm receives under the EU emissions trading scheme is dependent on its own report of its actual emissions. Consider n firms, indexed by i, which are distinguished by their emission values. The emission parameters θ^i are distributed over the interval $[0, \bar{\theta}]$ according to a probability density function $f:[0, \bar{\theta}] \to \mathbb{R}_+$. The government knows the probability distribution of the emission parameters, but not the realisations of specific firms. Firms have private information on their true emissions values θ^i and report possibly different values $\hat{\theta}^i$ to the government, in order to receive allowances. Since the

GHG allowances are grandfathered free of charge, firms maximize expected allowances.

The number of allowances for each country is determined by the Kyoto Protocol. Each country then determines how many of these allowances it will distribute to the companies that take part in the trading scheme. Therefore, the amount of aggregate allowances is previously determined by a political process. Let this predetermined bound be b. If the sum over all firms' initial allowances exceeds b, they can be adjusted respectively (as proposed by Harrison and Radov, 2002) via a compliance factor. The government has to rely on the firms' emission reports or perform costly audits.

The government introduces a *verification mechanism* $(\bar{e}(\cdot), a(\cdot), s(\cdot))$, which consists of the *allocation function*, determining the allocation before any compliance factor is applied:

$$\bar{e} : [0, \bar{\theta}] \to \mathbb{R}, \tag{1}$$

the *audit probability*

$$a : [0, \bar{\theta}] \to [0, 1], \tag{2}$$

and the *post-audit allocation function*, including a penalty for misreporting (and also called the *penalty function*)

$$s : [0, \bar{\theta}] \times [0, \bar{\theta}] \to \mathbb{R}. \tag{3}$$

Thus, a firm which reports θ^i receives initial allowances $\bar{e}(\theta^i)$ if it is not audited. It gets audited with probability $a(\theta^i)$ and then receives post-audit allowances $s(\theta^i | \theta^i)$. By assumption, the true parameter is discovered if a firm is audited.

The problem is comparable to tax evasion. There are, however, slight differences. While in the tax evasion context taxpayers have incentives to understate income values, they might prefer to overstate emission reports in this context. Another difference lies in the fact that there is an upper limit of allowances, whereas in the tax context, the government's objective might be the maximization of gross or net (expected) revenue, given a limited auditing budget (as in Border and Sobel, 1987; Reinganum and Wilde, 1985 and others). The competition for a fixed amount of allowances makes the problem similar to those in a rent-seeking context. However, it is not considered as such, since it is more reasonable that, due to the huge number of firms, they do not take the other firms' actions into consideration. The existence of the upper limit is treated as follows: expected allowances for firm i are given by:

$$\pi(\hat{\theta}^i \mid \theta^i) = (1 - a(\hat{\theta}^i))\overline{e}(\hat{\theta}^i) + a(\hat{\theta}^i)s(\hat{\theta}^i \mid \theta^i) \qquad (4)$$

A compliance factor δ is then determined such that $\delta \sum_{i=1}^{n} \pi(\hat{\theta}^i \mid \theta^i) = b$, i.e., to ensure that, given the signals of all firms, the allocations do not exceed the upper limit b. However, since firms do not take their rivals' actions into consideration, they take δ as a constant and neglect it in their optimization. From now on the index i is dropped.

From economic theory it is known that it is optimal to increase penalties for misreporting as far as possible and minimize the probability of costly auditing (cf. Becker, 1968). However, penalties that are too high are likely to be politically undesirable. Therefore, the penalty is given exogenously and regards that 'punishment should fit the crime.' In addition, there might be small, unintentional errors by the firm.[6] For the first part of the analysis, I therefore choose a post-audit allocation function which entails a reduction of the pre-audit allowances by the size of the difference between the true and the declared value:

$$s_{\overline{e}}(\hat{\theta} \mid \theta) = \overline{e}(\hat{\theta}) - \max\{0, \hat{\theta} - \theta\}. \qquad (5)$$

There is no reward for underreporting or truth-telling, an assumption that reflects discussions in the political context. Since it can be assumed that firms do not underreport, the analysis will concentrate on the case $\hat{\theta} > \theta$. The post-audit allocation function may become negative if, for example, the pre-audit allocations are relatively small and the declared value differs much from the true value.

The set F of feasible mechanisms is given by:

$$\overline{e}(\hat{\theta}) \geq 0 \quad \forall \hat{\theta}, \qquad (6)$$

$$\overline{e}(0) = 0, \qquad (7)$$

$$\pi(\hat{\theta} \mid 0) \leq 0 \quad \forall \hat{\theta}, \qquad (8)$$

$$F = \{(\overline{e}, a, s): \quad s = s_{\overline{e}} \text{ and } (6) - (8) \text{ hold}\}. \qquad (9)$$

Inequalities (6) ensure that the transfer of (pre-audit) allowances from the government to the firms is positive. If a firm reports zero emissions, it will not receive any allowances, which is given by equation (7). Equation (8) implies that firms do not expect to receive any allowances, if their true emission value is zero. (8) together with (2) leads to:

$$\overline{e}(\hat{\theta}) \leq \hat{\theta} \quad \forall \hat{\theta}, \tag{10}$$

This implies in particular the compactness of F, since $\hat{\theta} \in [0, \overline{\theta}]$.

All mechanisms in F are *direct revelation mechanisms*, since the message space equals the action space $[0, \overline{\theta}]$. A direct revelation mechanism is said to be *incentive compatible* if it is optimal for each firm to report its parameter truthfully. An equivalence to the revelation principle can be used to restrict the analysis to incentive compatible mechanisms:

Lemma 1. *Let* $(\overline{e}, a, s) \in F$ *be some mechanism. Then there exists an incentive-compatible direct revelation mechanism* $(\overline{e}', a', s') \in F$ *such that* $\overline{e}'$ *is nondecreasing and* $(\overline{e}', a', s')$ *replicates the equilibrium outcome arising from* $(\overline{e}, a, s)$.

The proof is given in the appendix. It is assumed that firms report their parameter truthfully if reporting truthfully is optimal.

A mechanism $(\overline{e}, a, s)$ is *efficient* if it is feasible and if there is no other feasible mechanism $(\overline{e}', a', s')$ with $\overline{e}'(\cdot) \geq \overline{e}(\cdot)$, $\overline{e}'(\cdot) \neq \overline{e}(\cdot)$ and $a'(\cdot) \leq a(\cdot)$, or $\overline{e}'(\cdot) \geq \overline{e}(\cdot)$, $a'(\cdot) \leq a(\cdot)$ and $a'(\cdot) \neq a(\cdot)$. That is, it is not feasible to allocate at least the same amount of allowances with less auditing.

Other things being equal, smaller audit probabilities are always preferred, as they imply lower audit costs. Now the main question is what the optimal nature of audit strategies is, and how these are related to the structure of allocations. An efficient mechanism exists, since the set of feasible mechanisms is compact.[7]

CHARACTERIZATION OF VERIFICATION MECHANISMS

Due to Lemma 1, the analysis can be restricted to incentive-compatible mechanisms $(\overline{e}, a, s) \in F$ with $\overline{e}$ nondecreasing:

$$\hat{\theta}^i > \hat{\theta}^j \longrightarrow \overline{e}(\hat{\theta}^i) \geq \overline{e}(\hat{\theta}^j), \tag{11}$$

$$\overline{e}(\theta) \geq \overline{e}(\hat{\theta}) - a(\hat{\theta})(\hat{\theta} - \theta). \tag{12}$$

$$F_I = \{(\overline{e}, a, s) : (\overline{e}, a, s) \in F \text{ and } (11) \text{ and } (12) \text{ hold.}\} \tag{13}$$

It can be seen that for nondecreasing allocation functions $\bar{e}$, it is a necessary condition for incentive compatibility that the audit probability $a(\hat{\theta})$ is not equal to zero for every reported $\hat{\theta}$.

The EU Directive (EU, 2003) requires a 'high degree of certainty' regarding the reports of the emission data. The following propositions answer the question of how precisely the audit probability should be fixed, and show in particular that *random auditing*, i.e., $a(\hat{\theta}) < 1$ for some $\hat{\theta}$, is preferable to *full verification*, i.e., $a(\hat{\theta}) = 1$ for all $\hat{\theta}$:

Proposition 1. *Let* $(\bar{e}, a, s) \in F_I$ *be an audit-efficient mechanism in* F_I. *The audit probability is determined by:*

$$a(\hat{\theta}) = \sup_{\theta < \hat{\theta}} \frac{\bar{e}(\hat{\theta}) - \bar{e}(\theta)}{\hat{\theta} - \theta} \quad \forall\, \hat{\theta}. \tag{14}$$

Proof: For any $(\bar{e}, a, s) \in F_I$, incentive compatibility is equivalent to:

$$a(\hat{\theta}) \geq \sup_{\theta < \hat{\theta}} \frac{\bar{e}(\hat{\theta}) - \bar{e}(\theta)}{\hat{\theta} - \theta} \quad \forall\, \hat{\theta}. \tag{15}$$

For $(\bar{e}, a, s) \in F_I$ efficient in F_I equality holds. If not, i.e., if:

$$a(\hat{\theta}) > \sup_{\theta < \hat{\theta}} \frac{\bar{e}(\hat{\theta}) - \bar{e}(\theta)}{\hat{\theta} - \theta},$$

it would be possible to decrease the audit probability without changing the allocations, which contradicts efficiency. ∎

For given allocation functions, the audit probability can be further specified:

Corollary 1. *Let* $(\bar{e}, a, s)$ *be an efficient mechanism in* F_I. *Then for concave allocation functions* $\bar{e}$, *the audit probability is nonincreasing in the reported parameter and equals average allocations. For convex functions* $\bar{e}$, *it is nondecreasing and equals marginal allocations. In particular, the audit probability is constant if the allocation function* $\bar{e}$ *is linear.*

Proof: Suppose $\bar{e}$ is *concave*. Then in (14) the supremum is attained at $\theta^i = 0$ and it holds:

$$a(\hat{\theta}) = \frac{\bar{e}(\hat{\theta})}{\hat{\theta}}. \tag{16}$$

Since $\bar{e}$ is concave, the average allocations are nonincreasing. If $\bar{e}$ is *convex*, the supremum is attained at $\theta = \hat{\theta}$:

$$a(\hat{\theta}) = \lim_{\theta \to \hat{\theta}, \theta < \hat{\theta}} \frac{\bar{e}(\hat{\theta}) - \bar{e}(\theta)}{\hat{\theta} - \theta} = D^- \bar{e}(\hat{\theta}). \qquad (17)$$

a $(\hat{\theta})$ equals the left-hand derivative. $\bar{e}$ is convex, therefore, $D^- \bar{e}$ is nondecreasing in the reported parameter and it holds that a is nondecreasing. The linear case follows directly. ∎

This result shows that the interplay between initial allocations and audit probabilities is dependent on the form of the allocation function. For a proportional allocation of allowances which can be observed in reality, all firms should be audited with the same probability. For concave allocation functions, the allowances will be allocated regressively. However, in order to obtain audit efficiency, the accompanying audit probabilities should be nonincreasing, which seems to be striking, since it means that companies which present higher emission reports are less likely to be audited. One has to consider, however, that the penalty and post-audit allocation function were chosen in a way that firms have incentives to announce their true values. This implies that a higher report does not stand for a bigger false report, but for a higher true value. A convex allocation function, on the other hand, corresponds to a progressive allocation of emission allowances, which might not be desirable. Audit efficiency then implies that a plant which reports higher values will be audited with a higher probability. From (17) and $a \in [0,1]$ it is immediately clear that only those mechanisms which have an allocation function with a first derivative between zero and one can be audit-efficient. The restriction is even stronger. From (15) and $\bar{e}$ convex, it follows that incentive compatibility is equivalent to a $(\hat{\theta}) \geq D^- \bar{e}$ $(\hat{\theta})$, which does not hold for first derivatives greater than one. Thus, mechanisms with $D^- \bar{e}$ $(\hat{\theta}) > 1$ are not element F_I.

The results depend on the underlying penalty function, not on the underlying density function.

Figure 8.1 above displays a concave and nondecreasing allocation function $\bar{e}$, corresponding to a regressive allocation of allowances. The audit probability is equal to the average allocation of allowances, as shown in (16). In the figure, it is given by the slope of the straight line through (0,0) and $[\hat{\theta}, \bar{e}, (\hat{\theta})]$.[8] A higher reported value leads to a lower audit probability. A scheme with a strictly concave allocation function $\bar{e}$ as displayed in Figure 8.1 involves random auditing, since the average allocations are strictly between zero and one.

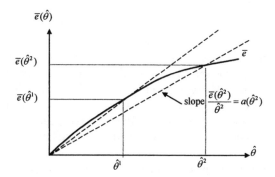

Figure 8.1   Relationship between allocation function and audit probability in efficient mechanisms

Corollary 2. *There is only one feasible allocation function $\overline{e}(\hat{\theta})$ for which full verification is audit efficient in F_I. This is the function $\overline{e}(\hat{\theta})=\hat{\theta}$.*

The proof is given in the appendix.

Up to now, the problem has been treated as analogous to tax evasion. However, as I mentioned earlier, there is a difference in the fact that there exists an upper limit of allowances, which results from the commitments under the Kyoto Protocol. Therefore, after the functional form of $\overline{e}$ has been fixed and after firms have submitted their reports, the government will choose the compliance factor δ which equates expected allocations to the given upper limit. This results in a change of the allocated pre- and post-audit allowances, but not in a change of the audit probability.

If the government, for example, expects firm data to be in a range that requires a compliance factor of approximately 0.9, and if it wishes to allocate the allowances proportionally, it should use a mechanism with allocation function $\overline{e}(\hat{\theta}) = 0.9\hat{\theta}$, audit probability $a(\hat{\theta}) = 0.9$ for each firm and the above described penalty. In this case, full verification will only be cost-efficient if the allocation function is $\overline{e}(\hat{\theta}) = \hat{\theta}$ and firms do not know that the compliance factor will be approximately 0.9. A system of allocation functions, audit probabilities and penalties is less cost-intensive than full verification. In addition, it provides more flexibility, since it also provides cost efficiency for concave allocation functions.

MAXIMUM PENALTY

Auditing of emission data, or of private information in general, is costly. It seems plausible, however, that rough estimates of the data can be obtained at low costs, whereas it will be very time and cost-consuming to receive the exact values. In this case, it may be desirable to allocate lump-sum fines, since this would allow for rough estimates. In the tax evasion literature, such as in Chander and Wilde (1998) or Mookherjee and Png (1989),[9] penalty functions which apply the maximum penalty if an incorrect report is detected are very often chosen.

I therefore introduce a similar penalty function in this context of verification of emissions data. However, it is not possible to fix a maximum penalty as easily as this can be done in the tax evasion context. Clearly, the assignment of zero allowances would not suffice, since this might reward lying if the true parameter is close to zero. Therefore, it is reasonable to choose a negative value as a penalty.

For $C \in \mathbb{R}_+$ the post-audit allocation function is then given by:

$$s_C(\hat{\theta} \mid \theta) = \begin{cases} -C & \forall\, \hat{\theta} > \theta \\ \overline{e}(\hat{\theta}) & \forall\, \hat{\theta} \leq \theta. \end{cases} \tag{18}$$

The set of feasible mechanisms is the same as before, differing only in the form of the post-audit allocation function:

$$F_c = \{(\overline{e}, a, s) : s = s_C \text{ and } (6)\text{-}(8) \text{ hold}\}, \tag{19}$$

and respectively:

$$F_{CI} = \{(\overline{e}, a, s) : (\overline{e}, a, s) \in F_c \text{ and } (11) \text{ and } (12) \text{ hold}\} \tag{20}$$

The efficient mechanisms in F_{CI} can then be characterized and efficiency can be guaranteed if the audit probability is chosen respectively.

Proposition 2. *A mechanism* $(\overline{e},\, a,\, s) \in F_{CI}$ *is efficient in* F_{CI} *iff* $\overline{e}$ *is nondecreasing and a is nondecreasing with:*

$$a(\hat{\theta}) = \frac{\overline{e}(\hat{\theta})}{\overline{e}(\hat{\theta}) + C}. \tag{21}$$

The proof is given in the appendix.

This result includes in particular that for any given C, every allocation function that fulfils the feasibility requirements can be used to set up an efficient mechanism by setting $s = s_c$ and the audit probability according to equation (21).

The audit probability is dependent on the strength of the penalty. The parameter C is given exogenously, fixed by a political or legal process. For high values of C, the audit probabilities will be low and *vice versa*. Firms reporting higher values will be audited with a higher probability. This is explained by the fact that the penalty for lying is independent of the deviation of the reported parameter from the true parameter. Therefore, it is likely that firms either report the truth or deviate strongly. Although the actual fine is the same, the expected punishment for a smaller deviation would then be smaller than the expected punishment for a high deviation, due to the lower audit probability.

It follows directly that the audit probability of an efficient mechanism is never equal to one, unless the post-audit allocation is zero for all reported values which are higher than the true values.

Corollary 3. *A mechanism $(\bar{e}, a, s) \in F_{CI}$ with $a(\cdot) \equiv 1$ is only efficient in F_{CI}, if $C = 0$.*

Full verification is only efficient for a mechanism with maximum penalty structure if the maximum penalty for an overstated report is set equal to zero. This may not be desirable. It may, however, be necessary if there is no possibility of allocating negative amounts of allowances, i.e., to actually make firms pay.

CONCLUSION

Under the EU emissions trading scheme, firms initially receive an amount of allowances that depends on their own reports of their current emissions. Since firms will receive more allowances the higher their baseline emissions are, strict verification of their reports is necessary. In view of the tight schedule for the allocation of allowances and of the costs for the verification of the data, however, it seems unlikely that all plants can be audited sufficiently thoroughly.

I have analyzed verification mechanisms which combine the allocation of initial allowances with an audit probability and penalties for misreporting. Two different penalty structures are considered: linear penalties, which depend on the degree of misreporting, and maximum penalties, which apply an exogenously determined maximum penalty to every false report.

If the linear penalty is chosen and combined with a proportional allocation of allowances, which actually is the case in most countries, it is efficient to audit every firm with the same probability, irrespective of its report. If allowances are allocated regressively, each firm should be audited with a probability equal to its average allocations in order to achieve efficiency. Audit probabilities are nonincreasing in this case. An efficient mechanism with a progressive allocation of initial allocations will have nondecreasing audit probabilities, which equal marginal allocations.

If maximum penalties are applied, the audit probability of an efficient mechanism is inversely related to this penalty and increasing in the reported emissions value. This holds for a regressive, as well as a progressive, allocation of initial allowances.

In addition, these proposed verification mechanisms have been compared to full verification, which provides a maximal level of security. However, full verification only leads to efficiency under very strong restrictions regarding the allocation function and proves to be inefficiently costly in general. Therefore, random auditing combined with penalties should be applied.

APPENDIX

Proof of Lemma 1:

Let $(\bar{e}, a, s) \in F$. Define $(\bar{e}', a', s')$ from $(\bar{e}, a, s)$ as follows:

$$\bar{e}'(\theta) := \bar{e}\big(\alpha(\theta)\big) - a\big(\alpha(\theta)\big)\big(\alpha(\theta) - \theta\big),$$

$$a'(\theta) := a\big(\alpha(\theta)\big),$$

$$s'(\theta) := s_{\bar{e}'}(\theta),$$

where $\alpha(\theta)$, $(\alpha(\theta) \geq \theta)$, is an optimal signal for a firm of type θ.

(a) $\bar{e}(\alpha(\theta)) - a(\alpha(\theta))(\alpha(\theta) - (\theta))$ is the expected allocation for a firm of type θ under the mechanism $(\bar{e}, a, s)$ if it pronounces $\alpha(\theta)$. This is not smaller than 0, since $\bar{e}(\theta) \geq 0$ for all θ and the firm can secure itself a punishment of 0 if it pronounces the true type θ. Therefore, $\bar{e}'(\theta) \geq 0$. $\bar{e}'(0) = 0$ since $\bar{e}(\alpha(0)) - a(\alpha(0))\alpha(0) = 0$ (from (8) and the optimality of $\alpha(\cdot)$). $\pi'(\theta|0) \leq 0$ holds trivially. Thus, $(\bar{e}', a', s') \in F$.

(b) $(\bar{e}, a, s)$ and $(\bar{e}', a', s')$ are obviously equivalent from the point of view of the planner and each firm, from the definition of $\alpha'(\theta)$ and since:

$$(1 - a(\alpha(\theta)))\bar{e}(\alpha(\theta)) + a(\alpha(\theta))(\bar{e}(\alpha(\theta)) - (\alpha(\theta) - \theta))$$

$$= \overline{e}(\alpha(\theta)) - a(\alpha(\theta))(\alpha(\theta) - \theta) = \overline{e}'(\theta) = (1 - a'(\theta))\overline{e}'(\theta) + a'(\theta)(\overline{e}'(\theta) - (\theta - \theta)).$$

(c) Show $\overline{e}'$ nondecreasing. Let $\theta^1 > \theta^2$ show that $\overline{e}'(\theta^1) \geq \overline{e}'(\theta^2)$:

$$\overline{e}'(\theta^2) = \overline{e}(\alpha(\theta^2)) - a(\alpha(\theta^2))(\alpha(\theta^2) - \theta^2) \leq \overline{e}(\alpha(\theta^2)) - a(\alpha(\theta^2))(\alpha(\theta^2) - \theta^1) \leq$$

$$\leq \overline{e}(\alpha(\theta^1)) - a(\alpha(\theta^1))(\alpha(\theta^1) - \theta^1) = \overline{e}'(\theta^1)$$

holds for both $\alpha(\theta^2) > \theta^1$ and $\alpha(\theta^2) \leq \theta^1$, since $\theta^1 > \theta^2$ and $\alpha(\theta^1)$ optimal for θ^1.

(d) In order to show that $(\overline{e}, a, s)$ is incentive-compatible, it is now sufficient to show that $\overline{e}'(\theta) \geq (1 - a'(\hat{\theta}))\overline{e}'(\hat{\theta}) + a'(\hat{\theta})[\overline{e}'(\hat{\theta}) - \max\{0, (\hat{\theta} - \theta)\}]$ for all $\hat{\theta} > \theta$, since from $\overline{e}'(\hat{\theta})$ nondecreasing, incentive compatibility holds true for all $\theta \leq \theta$. Let $\hat{\theta} > \theta$, then $\overline{e}'(\theta) \geq \overline{e}'(\hat{\theta}) - a'(\hat{\theta})(\hat{\theta} - \theta)$ is equivalent to:

$$\overline{e}(\alpha(\theta)) - a(\alpha(\theta))(\alpha(\theta) - \theta) \geq \overline{e}(\alpha(\hat{\theta})) - a(\alpha(\hat{\theta}))(\alpha(\hat{\theta}) - \hat{\theta}) - a(\alpha(\hat{\theta}))(\hat{\theta} - \theta)$$

$$= \overline{e}(\alpha(\hat{\theta})) - a(\alpha(\hat{\theta}))(\alpha(\hat{\theta}) - \theta),$$

which holds, since $\alpha(\theta)$ is an optimal signal for a firm of type θ. ∎

Proof of Corollary 2: From (14) and $a(\hat{\theta}) = 1$, it follows that:

$$1 = \sup_{\theta < \hat{\theta}} \frac{\overline{e}(\theta) - \overline{e}(\theta)}{\hat{\theta} - \theta} \quad \forall \hat{\theta}. \tag{22}$$

The equality holds for $\overline{e}(\hat{\theta}) = \hat{\theta}$. To show that this is the only function, let $\overline{e}(\hat{\theta}) \neq \hat{\theta}$ and $\hat{\theta}'$ the smallest value with $\overline{e}(\hat{\theta}') \neq \hat{\theta}'$, i.e., $\overline{e}(\theta) = \theta$ for all $\theta < \hat{\theta}'$ ($\hat{\theta}' > 0$, since $\overline{e}(0) = 0$). This includes:

$$\frac{\overline{e}(\hat{\theta}') - \overline{e}(\theta)}{\hat{\theta}' - \theta} < 1$$

for all $\theta < \hat{\theta}'$ and therefore:

$$\sup_{\theta < \hat{\theta}'} \frac{\overline{e}(\hat{\theta}') - \overline{e}(\theta)}{\hat{\theta}' - \theta} < 1,$$

which contradicts (22). ∎

Proof of Proposition 2:(a) $\bar{e}$ nondecreasing holds for all mechanisms $(\bar{e}, a, s)$ $\in F_{CI}$: for all $\hat{\theta} \leq \theta$ incentive compatibility is equivalent to $\bar{e}(\theta) \geq \bar{e}(\hat{\theta})$ for all θ, and for all $\hat{\theta} > \theta$ it is equivalent to:

$$\bar{e}(\theta) \geq (1 - a(\hat{\theta}))\bar{e}(\hat{\theta}) - a(\hat{\theta})C, \Leftrightarrow \bar{e}(\theta) \geq \bar{e}(\hat{\theta}) - a(\hat{\theta})[\bar{e}(\hat{\theta}) + C],$$

$$\Leftrightarrow a(\hat{\theta}) \geq \frac{\bar{e}(\hat{\theta}) - \bar{e}(\theta)}{\bar{e}(\hat{\theta}) + C}.$$

Since $a(\theta) \geq 0$, this implies $\bar{e}(\hat{\theta}) \geq \bar{e}(\theta)$.

(b) With similar reasoning as in the linear penalty case, it can be shown that efficiency implies:

$$a(\hat{\theta}) = \sup_{\theta < \hat{\theta}} \frac{\bar{e}(\hat{\theta}) - \bar{e}(\theta)}{\bar{e}(\hat{\theta}) + C}.$$

Independent of the form of the allocation function, the supremum is attained at $\theta = 0$, since $\bar{e}(0) = 0$ and $\bar{e}$ nondecreasing. Therefore, the audit probability of an efficient mechanism satisfies:

$$a(\hat{\theta}) = \frac{\bar{e}(\hat{\theta})}{\bar{e}(\hat{\theta}) + C}.$$

a nondecreasing follows then directly from $\bar{e}$ nondecreasing.

(c) To show that:

$$a(\hat{\theta}) = \frac{\bar{e}(\hat{\theta})}{\bar{e}(\hat{\theta}) + C}$$

is sufficient for efficiency, two cases are distinguished:

i) Let $(\bar{e}', a', s') \in F_{CI}$ with $\bar{e}'(\cdot) \geq \bar{e}(\cdot)$, $\bar{e}'(\cdot) \neq \bar{e}(\cdot)$.
From incentive compatibility it follows that:

$$a'(\hat{\theta}) \geq \sup_{\theta < \hat{\theta}} \frac{\bar{e}'(\hat{\theta}) - \bar{e}'(\theta)}{\bar{e}'(\hat{\theta}) + C}$$

Then there exists some $\hat{\theta} \in [0, \bar{\theta}]$ with:

$$a(\hat{\theta}) = \frac{\bar{e}(\hat{\theta})}{\bar{e}(\hat{\theta}) + C} < \frac{\bar{e}'(\hat{\theta})}{\bar{e}'(\hat{\theta}) + C} = \sup_{\theta < \hat{\theta}} \frac{\bar{e}'(\hat{\theta}) - \bar{e}'(\theta)}{\bar{e}'(\hat{\theta}) + C} \leq a'(\hat{\theta}).$$

Thus there is no $(\bar{e}', a', s') \in F_{CI}$ with $\bar{e}'(\cdot) \geq \bar{e}(\cdot)$, $\bar{e}'(\cdot) \neq \bar{e}(\cdot)$ and $a'(\cdot) \leq a(\cdot)$ for all $\hat{\theta} \in [0, \bar{\theta}]$.

ii) Suppose there is no $(\bar{e}', a', s') \in F_{CI}$ with $\bar{e}'(\cdot) \geq \bar{e}(\cdot)$, $\bar{e}'(\cdot) \neq \bar{e}(\cdot)$.

From incentive compatibility it follows that for the smallest possible a' belonging to $(\bar{e}', a', s') \in F_{CI}$, it has to hold that:

$$a'(\hat{\theta}) = \sup_{\theta < \hat{\theta}} \frac{\vec{e}'(\hat{\theta}) - \vec{e}'(\theta)}{\vec{e}'(\hat{\theta}) + C} = a(\hat{\theta}),$$

since $\bar{e}'(\hat{\theta}) = \bar{e}(\hat{\theta})$ and

$$\sup_{\theta < \hat{\theta}} \frac{\vec{e}'(\hat{\theta}) - \vec{e}'(\theta)}{\vec{e}'(\hat{\theta}) + C} = \frac{\vec{e}'(\hat{\theta})}{\vec{e}'(\hat{\theta}) + C}.$$

Thus there is also no $(\bar{e}', a', s') \in F_{CI}$ with $\bar{e}'(\cdot) \geq \bar{e}(\cdot)$ and $a'(\cdot) \leq a(\cdot)$, $a'(\cdot) \neq a(\cdot)$ for all $\hat{\theta} \in [0, \bar{\theta}]$. Therefore $(\bar{e}, a, s) \in F_{CI}$ is efficient in F_{CI}. ∎

NOTES

* Financial support from the Deutsche Forschungsgemeinschaft (German Research Foundation) is gratefully acknowledged. I thank Julia Angerhausen, Claudia Kriehn and Wolfram Richter for helpful comments and discussions.

1. Six GHGs are addressed in the Kyoto Protocol: carbon dioxide (CO_2), methane (CH_4), nitrous oxide (N_2O), hydrofluorocarbons (HFCs), perfluorocarbons (PFCs) and sulphur hexafluoride (SF_6). The gases are compared by transforming them into CO_2 equivalents.

2. One current example of the comparison of pollution permit markets and traditional standard regulations when emissions and costs of production and abatement are imperfectly observed is Montero (2005).

3. See Hahn and Hester (1989), Tietenberg *et al.* (1999) and others. The allocation of the initial allowances, however, results in this context from the commitments under the Kyoto Protocol and thus forms a different problem than the national allocation of allowances.

4. Boemare and Quirion (2002) provide an overview of implemented emission trading schemes, drawing amongst other aspects attention to permit allocation.

5. See Heyes (2000) for an overview over the extensive literature on this topic.

6. In the tax evasion context, Boadway and Sato (2000) explicitly allow for unintentional compliance errors by taxpayers in an otherwise similar model. They find, that in their setting, a reward for honest reporting has to be supplied and unintentional tax evasion has to be penalized in order to induce truth-telling. The solution proposed by Becker would then not be optimal, since this would entail too great costs in terms of social welfare, given that the innocent are punished.

7. Border and Sobel (1987) supply for a similar setting an example where an optimal solution does not exist when the feasible set is not compact. Mookherjee and Png (1989) show that, provided the agent has a minimal degree of risk aversion, optimal mechanisms exist even when the feasible set is not compact.

8. This result differs from the corresponding result in the tax evasion context, where the audit probability should be set equal to marginal payment rates.

9. Mookherjee and Png set up general conditions under which random auditing is optimal. They use a model with risk-averse agents and argue that in equilibrium the agent will report his income truthfully and that, therefore, the penalties for misreporting can be made as large as possible, to provide maximum incentives to report truthfully without affecting the agent's expected utility. Reinganum and Wilde (1985) use lump-sum taxes and fines for their compaison of audit cutoff versus random audit rules. Dye (1986) analyzes optimal monitoring policies in a general principal-agent framework with lump-sum payments.

REFERENCES

Becker, G.S. (1968), 'Crime and punishment: An economic approach', *Journal of Political Economy*, **76**, 169-217.

Boadway, R. and M. Sato (2000), 'The optimality of punishing only the innocent: The case of tax evasion', *International Tax and Public Finance*, **7**, 641-664.

Boemare, C. and P. Quirion (2002), *Implementing Greenhouse Gas Trading in Europe: Lessons from Economic Theory and International Experiences*, FEEM Nota di Lavoro 35/2002.

Border, K. and J. Sobel (1987), 'Samurai accountant: A theory of auditing and pluder', *Review of Economic Studies*, **54**, 525-540.

Center for Clean Air Policy (2001): *Study on the monitoring and measurement of greenhouse gas emissions at the plant level in the context of the Kyoto mechanisms, Final Report*, Center for Clean Air Policy/TNO/FIELD, Washington. (http://www.environmental-center.com/articles/article1290/finalreport 0110.pdf).

Chander, P. and L.L. Wilde (1998), 'A general characterization of optimal income tax enforcement', *Review of Economic Studies*, **65**, 165-83.

Dye, R. (1986), 'Optimal monitoring policies in agencies', *Rand Journal of Economics*, **17**, 339-350.

EU (2004), *Commission Decision of 29/01/04 establishing guidelines for the monitoring and reporting of greenhouse gas emissions pursuant to Directive 2003/87/EC of the European Parliament and of the Council*, European Commission, Brussels (http://www.environmentagency.gov.uk/commondata/105385/mrdecision0104_775902.pdf).

EU (2001), *Proposal for a Framework Directive for Greenhouse Gas Emissions Trading within the European Community*, COM (2001)581, European Commission, Brussels (http://europa.eu.int/eur-lex/en/com/pdf/2001/en_501PC0581.pdf).

Hahn, R.W. and G.L. Hester (1989), 'Marketable permits: Lessons for theory and practice', *Ecology Law Quarterly*, **16**, 361-406.

Harrison, D. and D.B. Radov (2002), *Evaluation of Alternative Initial Allocation Mechanisms in a European Union Greenhouse Gas Emissions Allowance Trading Scheme*, prepared for DG Environment, European Commission, Brussels. (http://europa.eu.int/comm/environment/climat/allocation_xsum.pdf).

Heyes, A.G. (2000), 'Implementing environmental regulation: enforcement and compliance', *Journal of Regulatory Economics*, **17**, 107-129.

Kaplow, L. and S. Shavell (1994), 'Optimal law enforcement with self-reporting of behavior', *Journal of Political Economy*, **102**(3), 583-606.

Montero, J.P. (2005), 'Pollution markets with imperfectly observed emissions', *RAND Journal of Economics*, **36**, 645-660.

Mookherjee, D. and I.P.L. Png (1989), 'Optimal Auditing, Insurance and Redistribution', *Quarterly Journal of Economics*, **104**, 399-415.

Reinganum, J.F. and L.L. Wilde (1985), 'Income tax compliance in a principal-agent framework', *Journal of Public Economics*, **26**, 1-18.

Tietenberg, T., M. Grubb, A. Michaelowa, B. Swift and Z.X. Zhang (1999), *International Rules for Greenhouse Gas Emissions Trading*, UNCTAD. (http://www.unctad.org/ghg/Publications/intl_rules.pdf).

UNFCCC (1997), *United Nations Framework Convention on Climate Change, Kyoto Protocol to the United Nations Framework Convention on Climate Change*, FCCC/CP/L.7/Add.1, Kyoto.

PART III

Advanced Issues in Climate Change and Energy
Policies

9. Induced technological change and slow energy capital stock turnover in an optimal CO_2 abatement model

Malte Schwoon and Richard S.J. Tol

INTRODUCTION

Scenarios of energy use and CO_2-emissions as well as the costs of limiting climate change depend crucially on technological change. Recently, energy modelers have started implementing research and development (R&D) and learning-by-doing (LBD) into energy/climate models so as to endogenize technological change. This turned out to have significant effects on the cost of stabilizing atmospheric CO_2-concentration and on the optimal timing of abatement efforts. The majority of models reveal drastic reductions in compliance costs, particularly in the case of LBD, as learning is basically for free (Azar and Dowlatabadi, 1999; Grubb et al., 2002; Löschel, 2002). Additionally, most LBD models suggest increased near term abatement to realize learning potentials (Rasmussen, 2001 and Manne and Richels, 2004 are exceptions). More diverse results can be found in the case of R&D. Goulder and Schneider (1999) and, more elaborately, Popp (2004), show that the impact of R&D on total abatement costs depends on assumptions about potential crowding out and spillovers between environmental and non-environmental R&D. Furthermore, including R&D seems to have a rather negligible effect on optimal timing decisions. Goulder and Mathai (2000) – from here on referred to as GM – show analytically that induced technological change necessarily reduces total costs of stabilizing the atmosphere at any given level. Moreover, initial abatement is always lower with R&D than without. On the other hand, the effect of LBD on initial abatement is ambiguous in theory, although a wide range of numerical simulations show that LBD has a strong increasing effect on early abatement.

In this chapter we will show that GM's result on the optimal timing of abatement to achieve lowest cost stabilization of atmospheric CO_2 concentration is biased by their assumption that there is no inertia in the energy system. GM neglect that there are barriers to the diffusion of new energy technologies. The most obvious barrier is the slow turnover time of energy generating and consuming capital stock, as shown in Figure 9.1. Thus, significant emission reductions in a short time imply expensive premature capital turnover costs.[1]

We include such inertia in the energy system by penalizing rapid changes in the level of abatement from one period to the next in the spirit of Ha-Duong *et al.* (1997).

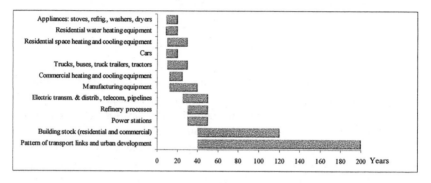

Source: IEA (2000), p. 43.

Figure 9.1   Typical time spans for capital-stock turnover

In contrast to GM's numerical simulations and the large majority of LBD studies to date, the model presented here shows a small, perhaps even negative influence of LBD on early abatement. On the other hand, the negative effect of R&D on initial abatement derived by GM turns out to be robust against the inclusion of inertia in the energy system. But in general, inertia seems to be a much more important determinant of the optimal abatement path than induced technological change (ITC) is. However, the reductions of the costs of stabilizing atmospheric CO_2-concentrations due to ITC predicted by GM are still in place.

In the next section, the analytical model of ITC via LBD or R&D and inertia is introduced. The third section is dedicated to numerical simulations. The chapter concludes with a discussion of the main policy implications and limitations of the model.

THE MODEL

Following GM's notation, $C(A_t, H_t, E_t^0)$ is the aggregate abatement cost function of a competitive economy, in which producers minimize costs. The costs include installation costs but not costs from increasing the speed of installation. A_t denotes the level of abatement at time t, H_t refers to the level of technology (or knowledge), and E_t^0 are the baseline emissions allowing the cost function to depend on the relative rather than the absolute level of abatement. Like GM, we assume that $C_A > 0$, $C_{AA} > 0$, $C_H < 0$, $C_{AH} < 0$, $C_E < 0$ and $C_{AE} < 0$.

We define costs of change as $C^R (R_t)$, where R_t is a variable reflecting the speed of change in abatement and is related to the difference between current and previous abatement. We accept some loss of generality by setting $C^R (R_t) = mR_t^2$, so that $C_R^R > 0$ and $C_{RR}^R > 0$. The parameter m is a measurement of the magnitude of inertia. Using the square of the rate of change ensures that not only rapid switches from low to high levels of abatement are expensive, but also rapid switches back, i.e., once our cars are running on hydrogen, it would be costly to move to carbon fuels again. Following Ha-Duong *et al.* (1997), we assume that direct abatement costs and change costs are separable, i.e., total abatement costs are $C^{tot} (A_t, H_t, E_t^0, R_t) = C(\cdot) + C^R(\cdot)$.

Technological change through R&D

In the R&D formulation of the model, the social planner has to choose an optimal abatement path together with optimal investments in R&D to accumulate knowledge. Adding the corresponding equations of motion for knowledge as well as atmospheric CO_2-concentrations and a concentration constraint, the planner faces the following intertemporal minimization problem:

$$\min_{A_t, I_t} \int_0^\infty \left(C(A_t, H_t, E_t^0) + mR_t^2 + p(I_t)I_t \right) e^{-rt}\, dt \tag{1}$$

$$\text{s.t.} \quad \dot{R}_t = -R_t + A_t - A_{t-1} \text{ for } t \geq 1 \tag{2}$$

$$\dot{S}_t = -\delta S_t + E_t^0 - A_t \tag{3}$$

$$\dot{H}_t = \alpha H_t + k\Psi(I_t, H_t) \tag{4}$$

$$R_0 = 0 \ , \ \dot{R}_0 = 0 \tag{5}$$

$$H_0, S_0 \text{ given and } S_t \leq \bar{S} \text{ for } t \geq T. \tag{6}$$

I_t is investment into R&D, $p\ (I_t)$ is the real price of a unit of research, r is the discount rate, S_t is atmospheric CO_2-concentration, δ is the natural rate of CO_2 uptake, α is the rate of autonomous technological change, $\Psi\ (\cdot)$ is the knowledge accumulation function, k is a parameter that defines the magnitude of ITC and $\bar{S}$ is the stabilization target to be met from year T onwards.

Expression (1) states that the planner minimizes the discounted sum of direct abatement costs, costs of change and investment costs from now on to the infinite future. Expression (2) defines that R changes with the difference between current and previous abatement. So R can be handled as a state variable that depends on the control variable and a lagged control variable, and therefore the system is solvable as a model with delayed response following Kamien and Schwartz (1991, Part II, Section 19).[2] The choice of this setup requires some discussion. A more intuitive approach might be to set $\dot{A}_t = R_t$ with R_t as a control variable and A_t as a state variable. But this approach was dismissed because distinct results concerning the behavior of the optimal path of R are difficult to obtain, even though the initial setup looks more concise. The reason is that results depend on the development of the shadow price associated with the change in abatement. And this development turns out to be ambiguous at every point in time. Moreover, the advantage of the current approach is that the GM model is fully nested (for $m = 0$).

The equation of motion for S (expression (3)) shows that CO_2-concentration declines through natural removal of atmospheric CO_2 and abatement, but increases through baseline emissions. According to expression (4), knowledge rises by the exogenous rate α plus induced knowledge accumulation. $\Psi\ (\cdot)$ is assumed to be positive and increasing in I_t with diminishing returns ($\Psi_I\ (\cdot) > 0, \Psi_{II}\ (\cdot) < 0$). Whether $\Psi_H\ (\cdot) > 0$ or $\Psi_H\ (\cdot) < 0$ depends on whether knowledge accumulation is characterized by 'standing on shoulders' or 'fishing out'. The relative importance of ITC compared to autonomous technological change is defined through k, which is used in the analysis as a switch to 'turn on ITC' through an increase from $k = 0$ to $k > 0$, allowing for comparative statics. The conditions subsumed in (5) take care of the starting point problem due to the lagged control variable in (2); and (6) states that CO_2-concentrations may not exceed the concentration target $\bar{S}$ from period T on.

For an analytically convenient solution of the model, we switch to a maximization problem, so that the current value Hamiltonian of the problem is:

$$\mathcal{H}_t = -C(A_t, H_t, E_t^0) - mR_t^2 - p(I_t)I_t$$

$$-\tau_t(-\delta S_t + E_t^u - A_t) + \mu_t(\alpha H_t + k\Psi(I_t, H_t)) + \lambda_t(-R_t + A_t - A_{t-1}) \quad \text{for } t < T,$$

where τ_t, μ_t and λ_t are the shadow values of S_t, H_t and R_t, respectively. Note that a negative τ_t is used to ensure a positive shadow value (i.e., the marginal benefit) of abatement. For $t \geq T$, the concentration constraint must be satisfied; hence, it is necessary to form the Lagrangian:

$$L_t = \mathcal{H}_t + \eta_t(\overline{S} - S_t).$$

Assuming an interior solution, and therefore ruling out negative abatement, together with costate equations, state equations and transversality conditions, a set of first-order conditions (FOCs) can be derived from the maximum principle. Since the negative of (1) and (3) as well as (4) are concave in A, R and H (assuming $\Psi_{HH}(\cdot) \leq 0$), the FOCs are also sufficient (Kamien and Schwartz, 1991, Part II, Section 3). The relevant necessary conditions that provide insights with respect to the optimal abatement path are:

$$\frac{\partial \mathcal{H}_t}{\partial A_t} + \frac{\partial \mathcal{H}_t}{\partial A_{t-1}}\bigg|_{t+1} = 0 \Leftrightarrow C_A - \lambda_t + \lambda_{t+1} = \tau_t \qquad \text{(FOC 1)}$$

$$\dot{\lambda}_t = r\lambda_t - \frac{\partial \mathcal{H}_t}{\partial R_t} - \frac{\partial \mathcal{H}_t}{\partial R_{t-1}}\bigg|_{t+1} \Leftrightarrow \dot{\lambda}_t = (1+r)\lambda_t + 2mR_t \qquad \text{(FOC 2)}$$

$$\text{and} \quad -\dot{\tau} = r(-\tau) - \frac{\partial L_t}{\partial S_t} \Leftrightarrow \dot{\tau}_t = (r+\delta)\tau_t - \begin{cases} 0, & \text{for } t < T \\ \eta_t, & \text{for } t \geq T. \end{cases} \qquad \text{(FOC 3)}$$

(FOC 1) states that in each period optimality requires the sum of the marginal costs of abatement and the value of a change in current abatement evaluated in the next period λ_{t+1} less the current value of a change in abatement λ_t to be equal to the shadow-costs of CO$_2$ emissions τ_t. (FOC 2) shows that λ_t is not necessarily strictly increasing or strictly decreasing over the whole time horizon, but depends on R_t. However, an acceleration of abatement activities implies an increase of the multiplier due to increasing marginal transition costs. To achieve a more intuitive expression of (FOC 1), a first step is solving the differential equation in (FOC 2) with respect to t, which gives:

$$\lambda_t = -2m \int_t^\infty R_s e^{-(1+r)(s-t)} \, ds .$$ (7)

Using (7) for λ_t and λ_{t+1} in (FOC 1) leads to:

$$C_A + 2m \left(\int_t^\infty R_s e^{-(1+r)(s-t)} \, ds - \int_{t+1}^\infty R_s e^{-(1+r)(s-t-1)} \, ds \right) = \tau_t .$$

Now, for incremental time steps, the difference in parentheses can be simplified to get:

$$C_A + 2m \left((1 - e^{1+r}) \int_t^\infty R_s e^{-(1+r)(s-t)} \, ds + e^{1+r} R_t \right) = \tau_t .$$ (FOC 1')

GM define τ_t as the optimal carbon tax being a pure Pigouvian tax that corrects the global externality associated with CO_2-emissions. Here, the optimal tax must equal the sum of direct marginal costs of abatement and marginal costs of the difference between current and future periods' discounted sum of the rates of abatement. The explanation for this is that, e.g., if abatement rises over time, an addition in current abatement increases the current rate (and therefore current costs of change) but reduces the rate in the next period. We call the net effect the costs of abatement shifting.

Optimality requires that the FOCs are fulfilled at every point in time. Therefore, we can answer the question of how the optimal abatement path reacts to the introduction of transition costs by differentiating (FOC 1') with respect to m and evaluating it at $m = 0$. Rearranging terms shows how optimal abatement changes in the initial period if m is altered:[3]

$$\frac{dA_0}{dm} = \frac{\dfrac{d\tau_0}{dm} - 2 \left((1 - e^{1+r}) \int_0^\infty R_s e^{-(1+r)s} \, ds \right)}{C_{AA}} .$$ (8)

A_0 tends upward, because first of all the denominator is positive by assumption. Then $d\tau_t / dm$ is clearly non-negative, because switching from $m=0$ to $m > 0$ imposes an additional cost to abatement and therefore increases the shadow-costs of emissions. The second term in the numerator is a measurement of the change in the costs of abatement shifting in the initial period. The integral is non-negative, because from abatement being non-negative, it follows that the sum of all possible decreases in abatement cannot exceed the sum of all increases, and furthermore, the increases must happen before the decreases and are therefore discounted less. Altogether, the change in the costs of abatement shifting is negative. Intuitively, if the penalty to change is increased, then the optimal abatement path will be characterized by lower changes. Now, the negative sign leads to an overall positive numerator. So we can conclude that introducing costs of change put an upward pressure on initial abatement. Therefore, the above representation of the impacts of inertia can be seen as an analytical underpinning of Ha-Duong *et al.*'s (1997) or Grubb's (1997) numerical analyses that higher inertia shifts optimal early abatement upward. From here on the term 'inertia-effect' is used to refer to this tendency.[4]

To shed some light on the impacts of ITC in the presence of inertia, we can differentiate (FOC 1') with respect to k and evaluate the result at $t = 0$ to obtain:

$$\frac{dA_0}{dk} = \frac{\dfrac{d\tau_0}{dk}}{C_{AA}}, \tag{9}$$

which is the same expression as in GM, indicating that initial abatement changes in the same direction as the initial shadow-cost. GM demonstrate that the shadow-cost-effect is less than or equal to zero, implying a negative impact of R&D on initial abatement (this could be called 'R&D-effect' as opposed to the upward inertia-effect). Note that (9) is independent of m, which is simply due to the fact that the rate of change in the initial period is zero by assumption. It can be shown that for $t > 0$, the impact of R&D on the optimal abatement path is decreasing in m. The reason is straightforward: A large share of change costs within the total costs makes the optimal path approach the smoothest path that satisfies the constraints. Thus, the larger m is, the more expensive it is to move away from the smoothest path by taking advantage of the R&D option. However, concerning initial abatement, we find that there is a downward R&D effect and upward inertia effect. Analytically, there is no way to gain any information about the relative magnitude of the two impacts, but as the simulations will show, it appears that for reasonable parameters the upward inertia-effect tends to outweigh the downward R&D-effect.

Technological change through LBD

In this section we alter the above model, such that technological change 'comes for free' just by learning from the act of abatement. So in this case, the social planner only chooses an optimal abatement path, i.e., the planner minimizes:

$$\min_{A_t} \int_0^\infty (C(A_t, H_t, E_t^0) + mR_t^2)e^{-rt}\, dt \qquad (10)$$

subject to the same constraints as above. Besides that, we now have $\Psi(A_t, H_t)$ with $\Psi_A > 0$, $\Psi_{AA} < 0$.[5] in the knowledge accumulation function. Solving the problem involves the same assumptions about the existence of an interior solution and necessary first-order conditions. It turns out that all FOCs are identical to the R&D case with the exception of (FOC 1'), which now is:

$$C_A - \mu_t k\Psi_A + 2m\left((1-e^{l+r})\int_t^\infty R_s e^{-(l+r)(s-t)}\, ds + e^{l+r}R_t\right) = \tau_t . \quad \text{(FOC 4)}$$

Here the shadow-value of abatement is equal to the marginal costs of abatement and change of abatement, reduced by the value of learning from abating. Applying the same strategy as above, we differentiate (FOC 4) with respect to m and evaluate it at $m = 0$ and $t = 0$ to get:

$$\frac{dA_0}{dm} = \frac{\dfrac{d\tau_0}{dm} - 2\left((1-e^{l+r})\int_0^\infty R_s e^{-(l+r)s}\, ds\right) - k\Psi_A \dfrac{d\mu_0}{dm}}{C_{AA} + \mu_0 k\Psi_{AA}}, \qquad (11)$$

which determines the impact of a change in m on the optimal initial abatement.

Note that without LBD, i.e., $k = 0$, expression (11) is reduced to expression (8) of the R&D case and A_0 tends upward if m increases. But if $k > 0$, things get unclear. Since knowledge lowers abatement costs, the shadow value of knowledge μ_0 cannot be negative. Thus, by assuming decreasing returns in knowledge accumulation, the second term in the denominator is negative, so that the sign of the ratio is already ambiguous. Furthermore, it turns out to be impossible to handle the behavior of the third term in the numerator, which adds even more uncertainty.

Hence, in the presence of LBD, the impact of inertia on initial abatement is ambiguous. This might be a surprising result at first glance. Optimal initial abatement tends upward in the R&D specification; so one might expect such a consequence all the more if it is possible to gain learning effects from early abatement. But there is uncertainty about the impact of LBD itself as it is shown now. Following the same steps as in the R&D specification, we can derive:

$$\frac{dA_0}{dk} = \frac{\dfrac{d\pi_0}{dk} + \mu_0 \Psi_A}{C_{AA}},$$ (12)

for the impact of LBD on initial abatement, which is the same expression as in GM. There is a negative shadow-cost effect in the numerator as opposed to a positive LBD-effect (the product of the shadow value of knowledge and the marginal increase in knowledge from LBD, which are both positive). Actually, this effect is responsible for the positive effect of LBD on initial abatement in virtually all of GM's numerical simulations. But the numerical simulations will show that there is a strong upward inertia-effect that gives very little room to additional cost-effective increases through LBD.

NUMERICAL SIMULATIONS

Before we present the results of the numerical simulations, we briefly discuss the calibration of the model. Baseline emissions, the functional forms and main parameter values with respect to direct abatement costs, knowledge accumulation and CO_2 accumulation chosen for the simulations are as in GM.[6]

The costs of change evolve according to:

$$C^R(\cdot) = m^{\alpha_{c3}} \left| R_t \right|^{\alpha_{c4}},$$ (13)

Now we use $R_t = A_t - A_{t-1}$, which implies $\dot{R}_t = -R_t + A_{t+1} - A_t$. This has previously been avoided to achieve an analytically more convenient lag-structure instead of a lead-structure. Numerically this is irrelevant and has the advantage of gaining one more observation in the very beginning, which is of particular interest. The exponents α_{c3} and α_{c4} in equation (13) are set equal to 2 in the central case.[7] This formulation of change costs follows Ha-Duong *et al.* (1997), who claim that results from several modeling studies of capital stock turnover justify a nonlinear increase in costs. They interpret m as the characteristic timescale of turnover in the global energy system. From the half-life of capital for a depreciation rate of 4 percent, they derive $m = 20 \approx \ln 2/0.04$ years as a lower bound for inertia, representing a realistic assumption for the average rate of replacement of appliances and cars. To incorporate typical timescales of the turnover in energy systems and additional sources of inertia, Ha-Duong *et al.* (1997) also analyze a value of $m = 50$. In order not to overstate the impact of inertia, in this model a rather low end value of 30 is used for the central case.

Splitting up abatement costs into direct and change costs requires a recalibration. Thus, we scaled total abatement costs, such that the sum of total discounted costs (without ITC) including costs of change equals the sum in the GM central case. Additionally, with $m = 30$, the share of the sum of discounted transition costs within total costs over the next century is approximately 33 percent. For $m = 20$ and $m = 50$, the according percentages are 18 percent and 55 percent, respectively, which are lower but basically in line with the values of Ha-Duong *et al.* (1997), who report 31 percent and 71 percent. The model is solved in five-year steps from 2000 until 2300. In the last period, steady state conditions require abatement and CO_2-concentrations to remain constant, to eliminate any unsmooth behavior at the very end.

The impact of R&D in the presence of inertia

The results of the numerical simulations gain their significance from a comparison with the GM results. By calibration the two base case scenarios (the one with inertia and the original GM case) without ITC have the same sum of total discounted costs. Costs of change in the inertia case account for 33 percent of the total costs from 2000–2100, but are responsible for only some 10 percent of the total costs over the whole simulated time span. Moreover, the overall abatement profiles of the two cases hardly differ in the overall behavior of the optimal abatement path. So it can be concluded that the two cases are sufficiently comparable.

We want to get an impression of how the impact of R&D on initial abatement is changed in the presence of inertia. Figure 9.2 shows the percentage reduction of initial abatement due to R&D. Recall that the analytical results unambiguously predicted a decrease of initial abatement. In addition to the central parameterization case, Figure 9.2 includes the results for different assumptions regarding the discount rate r and the stabilization target S (formerly denoted as $\bar{S}$). GM use a Cobb-Douglas knowledge production function $\Psi\ (I_t,\ H_t\ )=M_\Psi\ I_t^\gamma\ H_t^\varphi$ with central case values $M_\Psi = 0.0022$ and $\varphi = \gamma = 0.5$.[8] Varying the parameters of the function implies different notions of knowledge accumulation. M_Ψ scales the magnitude of

ITC, $\varphi = 0.75$ implies stronger benefits from previous knowledge or oppositely 'fishing out' ($\varphi = -0.5$). Moreover, we allow for higher autonomous technological change ($\alpha = 0.01$, compared to 0.005 in the central case), and finally no autonomous technological change at all ($\alpha = 0$).

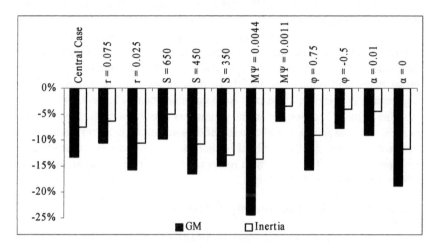

Figure 9.2 Change of optimal initial abatement through R&D (R&D relative to non-R&D)

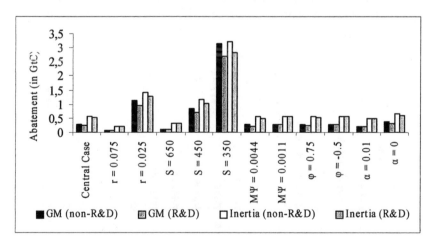

Figure 9.3 Abatement in 2000

By comparing the two central cases, it can be seen that the reduction is cut by half in the inertia-model compared to the GM model. Similar drastic declines in the impact of R&D are observable for alterations in the knowledge parameters. Reductions in the effect of R&D are still noticeable for different values of the discount rate and the stabilization target, with the only exception that for the extremely ambitious $S = 350$ ppmv target, R&D remains similarly powerful. The reason is that R&D only affects direct abatement costs. But a 350 ppmv target requires high immediate abatement, implying a high share of direct costs and a rather low share of change costs. Thus, the R&D-effect on initial abatement remains powerful. To provide some more insight into the importance of inertia with respect to initial abatement, Figure 9.3 shows the absolute values.[9] Independent of the parameters chosen (with the $S = 350$ case being somewhat aside), initial abatement depends much more on the question of whether or not the 'true' model is characterized by inertia rather than the presence of R&D. Therefore, with respect to the optimal timing of abatement efforts, inertia seems to be crucial.

The impact of LBD in the presence of inertia

GM's numerical simulations in the case of LBD show significant increases in initial abatement for almost all parameterizations (Figure 9.4). But in the presence of inertia, the upward LBD-effect and the downward shadow-cost-effect seem to balance each other. For $r = 0.025$, $M_\psi = 0.0044$, and $\alpha = 0$, the sign of the impact actually switches, and LBD lowers initial abatement.

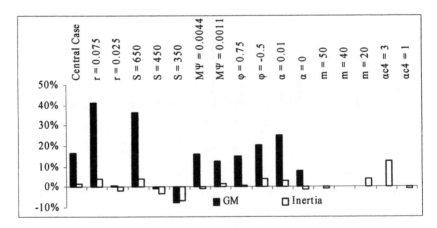

Figure 9.4   Change of optimal initial abatement through LBD (LBD relative to non-LBD)

Figure 9.4 also includes the change of initial abatement under different assumptions for the exponent of the rate of change (α_{c4}) and for the characteristic timescale of the energy system (m).[10] Decreasing the exponent implies that small rates are relatively expensive and *vice versa*. Therefore, Figure 9.4 illustrates that the costlier the changes are, the lower the impact of LBD is, and LBD can even be negative. Recall from the discussion of the analytical analysis that GM could not rule out a negative impact of LBD. But within their numerical results, the $S = 350$ ppmv case, being the only one with a negative impact, looks like an 'outlier' for a rather unrealistic scenario.

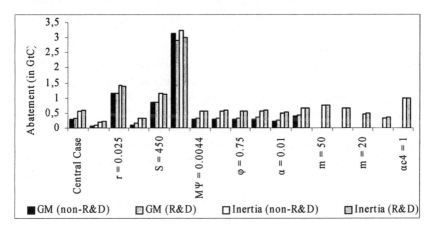

Figure 9.5   Abatement in 2000

So why does inertia almost eliminate the generally substantial positive effect on initial abatement, also contradicting Ha-Duong *et al.*'s (1997) claim that, within their analysis of inertia, they have probably underestimated the value of early abatement as a result of neglecting the effects of LBD. A straightforward reason can be derived from Figure 9.5, which goes back to the absolute values for initial abatement. Again, inertia shifts initial abatement upwards. Comparing the absolute values in Figure 9.5 with the relative changes in Figure 9.4, it becomes obvious that LBD has a decreasing impact exactly in those cases where initial abatement is already rather high. Then, due to decreasing returns in knowledge accumulation, the gains from additional early abatement to induce long term cost reductions are low. But on the other hand, the option to defer some more abatement efforts to the future, when they will be much cheaper because of discounting and 'free' knowledge enhancements from LBD, becomes much more valuable. Consequently, this latter cost-shifting potential of LBD outweighs the former cost reduction potential.

Reduction of stabilization costs

There is general agreement on the fact that ITC lowers the overall costs of stabilizing atmospheric CO_2-concentration. In our model, the present net value of total abatement costs is about \$400 billion without ITC. Without inertia, R&D lowers these costs by 23 percent in the central case. As R&D does not affect change costs, cost reductions are restricted to the direct costs of change, but even in rather inert systems ($m = 50$), costs are still reduced by some 17 percent. Thus, the cost-reducing effect of R&D seems to work as usual, so to say, but is limited to the fraction of direct costs. This indicates that declines in costs from R&D are due mostly to the long-term decrease in unit costs and only to a lesser extent by the option of lower early abatement.

Cost reductions from LBD, which are 'for free' as they do not require any investments in knowledge, are even more pronounced. For the central case, LBD decreases total costs to only \$270 billion, which is at the very low end of the range of most cost estimates.[11] Inertia limits reductions (by construction) to the fractions of direct abatement costs but, as in the R&D case, the major cost-decreasing effect of LBD is related to the reduction of future unit costs of abatement rather than timing, and therefore works independently of the assumed inertia.[12]

CONCLUSION AND DISCUSSION

This study investigates how inertia based on slow energy capital stock turnover alters the impact of R&D and LBD on optimal timing of abatement and explores results with a numerical model. For today's policy considerations, near-term costs and therefore optimal near term abatement requirements are particularly important. Analytically, we find that inertia shifts up initial abatement in the case of R&D and has an ambiguous impact in the case of LBD. The numerical model reveals that in both settings the level of inertia assumed determines early abatement to a much higher degree than ITC does. In the central case considered, inertia implies relatively high abatement over the next few decades. Although, analytically, R&D always reduces optimal initial abatement even in the presence of inertia, the numerical model shows that this effect is small, because at early stages the relatively high share of transition costs dominates optimal timing.

In the case of LBD, the model challenges the notion that potential gains from learning justify higher early abatement. Two interrelated effects are responsible for the fact that the model shows lower early abatement for parameter values that are not unrealistic. As inertia already implies relatively high initial abatement, additional upward pressure from LBD is limited because of decreasing returns from learning. Therefore, if learning potentials are sufficiently exhausted, optimality calls for a deferral of some abatement efforts as future actions are less costly due to the very act of learning; this is enhanced by discounting.

Earlier studies showed that ITC decreases the total costs of stabilizing CO$_2$ concentrations significantly. We find that even though ITC is assumed to have no decreasing impact on the cost associated with inertia, stabilization costs are still substantially lower in the presence of ITC. The reason is that the cost-decreasing effects of ITC are due to the reduction of future unit costs rather than to different timing of abatement.

There are several limitations to the present model. The empirical base for the parameters that describe the endogenous change of technology is weak. Moreover, the stabilization target is assumed to be known, while uncertainty about the target is addressed only *via* sensitivity analysis. The implications of uncertainty in optimal timing of abatement are not clear-cut. There might be benefits from waiting given the sunk cost character of abatement (Pindyck, 2000, 2002), but there are also irreversibilities in the climate calling for early action (Kolstad, 1996). Finally, the high level of aggregation necessary for analytical traceability is another drawback. A promising compromise would be to break down the economy into two sectors, as in Lecocq *et al.* (1998). The two sectors would differ in inertia, and therefore also in the effects of R&D and LBD. Alternatively, instead of two sectors, two countries which have different diffusion rates of energy-efficient technologies and have to achieve a common emission reduction target could be modeled. Since distributional effects are likely to have an important impact on climate targets and policies to achieve them, such a setting may provide valuable insights into burden sharing.

NOTES

1. For a discussion of other diffusion barriers, see, e.g., Sathaye *et al.* (2001), Ha-Doung *et al.* (1997) or Jaffe *et al.* (2002). The presence of what is called socio-economic inertia in the energy system implies that premature capital turnover costs are a low-end estimate of the total costs of accelerating the application of energy-efficient technology.

2. El-Hodiri *et al.* (1972) develop a growth model with lags in the employment of the optimal capital stock and Mann (1975) models the delay between advertising expenditures and their impact on sales. For an application of the technique in the context of resource economics, see Wilman and Mahendrarajah (2002).

3. For details on the derivations of the analytical results in this section, see Schwoon and Tol (2004).

4. It seems necessary to point out that the analytical result is only valid in the vicinity of $m = 0$, but simulations do not provide any indication not to believe that it also holds for other values of m.

5. The assumption of decreasing returns of learning is motivated by the empirical observation of experience curves. See, e.g., Dutton and Thomas (1984), Argote and Epple (1990) or Wene (2000) in the context of carbon-free energy sources.

6. We actually used a slightly different baseline emission path (basically following IPCC IS92a (Leggett *et al.*, 1992)), so for comparison of the results we recomputed GM's results.

7. For the sake of simplicity, m is used instead of m^2 in the analytical part. Furthermore, in the simulations, the employment of the absolute value of R_t makes it possible to do sensitivity analyses with respect to the exponent α_{c4}.

8. The same function is applied in the LBD case with I being replaced by A.

9. Information about the initial period also implies similar optimal behavior in the following periods, since almost all model runs show very smooth behavior at the beginning. So it is not the aim to say that optimal abatement in the year 2000 should have been, e.g., 0.5 GtC. It should rather be interpreted as 'relatively high' or 'relatively low' abatement within the next, say, 20 years, because comparable figures could be drawn until 2020.

10. Since the inertia model is calibrated for $\alpha_{c4} = 2$ and $m = 30$, there are no corresponding cases of the GM model. Note that for $m = 40$, LBD-effect and shadow-cost effect seem to balance exactly.

11. Recent studies such as van der Zwaan and Gerlagh's (2004) show similarly low costs.

12. Our result is consistent with Rasmussen (2001), who observes that welfare effects are not significantly altered if costs of capital adjustment are changed. Given the similarity of the findings, it would be interesting to see whether Rasmussen's (2001) model could confirm the responsiveness of initial abatement with respect to turnover costs, given the fact that he also observes a slight decline in initial abatement due to LBD.

REFERENCES

Azar, C. and H. Dowlatabadi (1999), 'A review of technical change in assessment of climate policy', *Annual Review of Energy and the Environment*, **24**, 513-544.

Argote, L. and D. Epple (1990), 'Learning curves in manufacturing', *Science*, **247**, 920-924.

Dutton, J.M. and A. Thomas (1984), 'Treating progress functions as a managerial opportunity', *Academy of Management Review*, **9** (2), 235-247.

El-Hodiri, M.A., E. Loehman, and A. Whinston (1972), 'An optimal growth model with time lags', *Econometrica*, **40** (6), 1137-1147.

Goulder, L.H. and K. Mathai (2000), 'Optimal CO$_2$ abatement in the presence of induced technological change', *Journal of Environmental Economics and Management*, **39**, 1-38.

Goulder, L.H. and S.H. Schneider (1999), 'Induced technological change and the attractiveness of CO$_2$ abatement policies', *Resource and Energy Economics*, **21**, 211-253.

Grubb, M., J. Köhler and D. Anderson (2002), 'Induced technical change in energy and environmental modeling, analytic approaches and policy implications', *Annual Review of Energy and the Environment*, **27**, 271-308.

Grubb, M. (1997), 'Technologies, energy systems and the timing of CO$_2$ emissions abatement', *Energy Policy*, **25** (2), 159-172.

Ha-Duong, M., M.J. Grubb and J.C. Hourcade (1997), 'Influence of socioeconomic inertia and uncertainty on optimal CO$_2$-emission abatement', *Nature*, **390**, 270-173.

International Energy Agency, IEA (2000), *World Energy Outlook 2000*, Paris: OECD Publications.

Jaffe, A.B., R.G. Newell, and R.N. Stavins (2002), 'Environmental policy and technological change', *Environmental and Resource Economics*, **22**, 41-69.

Kamien, M.I. and N.L. Schwartz (1991), *Dynamic Optimization – The Calculus of Variations and Optimal Control in Economics and Management*, 2nd Edition, New York: North-Holland.

Kolstad, C.D. (1996), 'Learning and stock effects in environmental regulation: The case of greenhouse gas emissions', *Journal of Environmental Economics and Management*, **31**, 1-18.

Lecocq, F., J.C. Hourcade and H. Ha-Duong (1998), 'Decision making under uncertainty and inertia constraints, sectoral implications of the when flexibility', *Energy Economics*, **20**, 539-555.

Leggett, J.A., W.J. Pepper and R.J. Swart (1992), 'Emission scenarios for the IPCC: An update', in J.T. Houghton, B.A. Callander and S.K. Varney (eds), Climate Change 1992, The Supplementary Report to the IPCC Scientific Assessment, Cambridge and New York: Cambridge University Press, 69-95.

Löschel, A. (2002), 'Technological change in economic models of environmental policy: a survey', *Ecological Economics*, **43**, 105-126.

Mann, D.H. (1975), 'Optimal Theoretic Advertising Stock Models: A generalization incorporating the effects of delayed response from promotional expenditure', *Management Science*, **21** (7), 823-832.

Manne, A.S. and R. Richels (2004), 'The Impact of learning-by-doing on the timing and costs of CO_2 abatement', *Energy Economics*, **26**, 603-619.

Pindyck, R.S. (2000), 'Irreversibilities and the timing of environmental policy', *Resource and Energy Economics*, **22**, 233-259.

Pindyck, R.S. (2002), 'Optimal timing problems in environmental economics', *Journal of Economic Dynamics and Control*, **26**, 1677-1697.

Popp, D. (2004) 'ENTICE: endogenous technological change in the DICE model of global warming', *Journal of Environmental Economics and Management*, **48**, 742-768.

Rasmussen, T.N. (2001), 'CO_2 abatement policy with learning-by-doing in renewable energy', *Resource and Energy Economics*, **23**, 297-325.

Sathaye, J., D. Bouille, D. Biswas, P. Crabbe, L. Geng, D. Hall, H. Imura, A. Jaffe, L. Michaelis, G. Peszko, A. Verbruggen, E. Worrell and F. Yamba (2001), 'Barriers, Opportunities, and Market Potential of Technologies and Practices', in B. Metz, O. Davidson, R. Swart and J. Pan (eds), *Climate Change 2001: Mitigation. Contribution of Working Group III to the Third Assessment Report of the Intergovernmental Panel on Climate Change*, 345-398.

Schwoon, M. and R.S.J. Tol (2004), 'Optimal CO_2-abatement with socio-economic inertia and induced technological change', FNU Working Paper #37, Centre for Marine and Climate Research, Hamburg/Germany.

van der Zwaan, B.C.C. and R. Gerlagh (2004), 'A sensitivity analysis of timing and costs of greenhouse gas emission reductions under learning effects and niche markets', *Climatic Change*, **65**: 39-71.

Wene, C.O. (2000), *Experience Curves for Energy Technology Policy*, International Energy Agency. Paris: OECD Publications.

Wilman, E.A. and M.S. Mahendrarajah (2002), 'Carbon offsets', *Land Economics*, **78** (3), 405-416.

10. Indeterminacy and optimal environmental public policies in an endogenous growth model

Rafaela Pérez and Jesús Ruiz

INTRODUCTION

Nowadays there are legal standards determining socially acceptable levels of pollutants in the environment. However, a producer has an incentive to shirk pollution control. Since a producer's profits come from a market price that does not reflect society's preferences for environmental protection, the producer has no economic incentive to supply the level of pollution control that society wants. If the market is not sending the correct signal to the producer about the socially optimal level of pollution control, a regulator can implement economic incentives that increase the cost of shirking pollution control. For example, the government could tax the polluting productive activities. Nevertheless, a coordination failure between private and government sectors could arise because the choice of tax policy affects private agents' decisions and, at the same time, government expenditures could yield external benefits to private agents. As a consequence, the environmental policy implemented by government may not yield the effects on pollution levels that society wants. From a theoretical point of view, the existence of coordination failure can be explained through the indeterminacy of equilibria issues.

This chapter studies, in a second-best framework, the dynamic properties of a general equilibrium model of endogenous growth with environmental externalities. In particular, we study the conditions for global and local indeterminacy, and their implications for the governmental control of pollution.

General equilibrium models that display indeterminacy have been a focus of attention in recent years. Global indeterminacy implies multiple balanced growth paths (BGP) along which the economy can persistently grow in the long run. On the other hand, local indeterminacy implies that there will be multiple paths converging to a given steady state.[1]

Characteristics that produce local indeterminacy of equilibria in one or multi-sector real business cycle models or in endogenous growth models have been widely studied. The seminal chapters (Benhabib and Farmer, 1994, and Farmer and Guo, 1994, among others) have been criticized because, to produce multiple equilibria, they require larger returns to scale than observed in actual data (see Aiyagari, 1995). However, Fernández, Novales and Ruiz (2003) obtain multiple equilibria under constant returns to scale and with endogenous government expenditures included in the utility function.

More recent works have described situations in which increasing returns needed to produce indeterminacy are lower than initially thought (see Benhabib and Farmer, 1996; Perli, 1998; Benhabib and Nishimura, 1998 or Wen, 1998), but the elasticity of intertemporal substitution needed is often too large, relative to values considered in literature on the real business cycle (Weder, 1998 and Bennett and Farmer, 2000, among others). The same criticism applies to existing endogenous growth models producing indeterminacy (see Benhabib and Perli, 1994, and Xie, 1994). One exception is Wen (1998), whose model can generate local indeterminacy with low increasing returns and smooth consumption.

All these previous chapters study indeterminacy for the decentralized solution. However, we study indeterminacy conditions in the Ramsey-type fiscal policy problem. In our analysis, we follow the methodology used by Park and Philippopoulos (2004), who study indeterminacy in an economy without pollution, very different from ours, in which capital taxes are used to finance public production and consumption services.

In our chapter, the economy includes the following features: i) pollution is a side-effect of productive activity and enters negatively in the utility function of agents (Smulders and Gradus, 1996, Ligthart and van der Ploeg, 1994, or Bovenberg and de Mooij, 1997,[2] are some examples), who negatively value the effects of pollution, for example, the possibility of climate change due to global warming, the damage of pollution on human health or pollution of rivers and lakes; ii) the government provides two types of services: first, those that increase labor efficiency in the productive sector and, second, those that improve the quality of the environment ('abatement activities'); iii) public expenditures are financed by a proportional tax charged on firms' revenues (the firms finance the abatement activities because they receive the benefits from the improvement in labor productivity and because they also perform the polluting productive activity).

Under this framework, the government at all times chooses the optimal composition of public spending, together with the optimal tax rate, taking into account the decentralized competitive equilibrium. We study how second-best economic policy influences the dynamics of growth, a task that has not been previously undertaken in environmental economics literature. We show that when we include pollution in the utility function together with public abatement activities and public productive services, the introduction of endogenously chosen economic policy causes a problem of expectations coordination: although the decentralized solution is determinate (globally and locally), we find that the 'second-best' solution can be globally and locally indeterminate; that is, there can be two steady states (global indeterminacy), both of which can be locally indeterminate (that is, there exist multiple converging paths to each state). This result is due to the interaction between the environmental externality in the utility function and the externality of public services in production: if only one externality is considered, the Ramsey-type equilibrium is determinate.

The implication for economic policy behind global indeterminacy is that two economies with the same fundamentals could choose different optimal policies (taxes and spending composition) that would yield different optimal paths of pollution.

On the other hand, local indeterminacy implies that there are multiple converging paths towards a given balanced growth path (BGP). The main implication for environmental policy is that in situations of indeterminacy, public policies are insufficient to drive the economy towards low pollution levels during the transition towards the long-run equilibrium.

Hence, local indeterminacy is able to explain, for example, how two countries with similar fundamentals might over time display different relationships between degree of development and environmental quality (or pollution). This relationship is known as the Environmental Kuznets Curve (EKC) and the existence of local indeterminacy could be a possible explanation for the difficulty of finding a clear empirical regularity supporting the EKC hypothesis (see De Bruyn and Heintz, 1999). In other words, the past pollution experience of a developed country does not guarantee that a developing country will follow the same path over time, even if the initial conditions and the environmental policies are similar.

THE ECONOMY

We consider a closed economy that displays endogenous growth with constant population normalized to one. In this section we set up the decentralized equilibrium. The private sector consists of a representative household and a representative firm, both competitive. Pollution appears as a production externality and has a negative effect on the household utility. The government sector taxes the representative firm (which pollutes) and uses the revenues to finance two types of government spending: i) abatement activities that improve environmental quality and ii) spending that increases labor efficiency in the productive sector.

The household

The representative household maximizes intertemporal utility:

$$\int_0^\infty e^{-pt} [\ln c - \eta \ln P] \, dt, \tag{1}$$

where c is consumption *per-capita*, P is the flow of pollution, the parameter $\rho > 0$ is the rate of time preference and $\eta > 0$ is the weight of pollution in utility. To simplify notation, time dependence of all variables is suppressed.

The household consumes and accumulates capital, and receives income from labor (which is offered in an inelastic manner) and from capital renting to firms:

$$c + \dot{k} = rk + w, \tag{2}$$

where r is the market return to capital and w the wage rate. A dot over a variable denotes time derivative and the initial stock $k_0 > 0$ is given.

The representative household is competitive, taking prices and pollution as given. The first-order conditions are the standard Euler condition:

$$\frac{\dot{c}}{c} = r - \rho, \tag{3}$$

apart from the budget constraint and the transversality condition $\lim_{t \to \infty} e^{-pt} \lambda k = 0$.

The firm

The representative competitive firm takes prices and public variables (taxes and government expenditures) as exogenously determined. It maximizes profits, $\pi = (1-\tau) Y - rK - wL$ subject to the technology constraint:

$Y = A (gL)^{1-\alpha} K^{\alpha}$, with $0 < \alpha < 1$, where τ is the tax rate on the firm's revenues, Y, g, K and L are, respectively, aggregate output, flow of public investment (government spending that increases private labor productivity), private capital and labor.

Technology can also be expressed in per capita terms or per labor units: $y = Ag^{1-\alpha} k^{\alpha}$, and hence, it is assumed that public productive services are not subject to congestion.

The first-order conditions for capital and labor are as usual:

$$r = (1-\tau)\alpha\frac{y}{k},$$ (4)

$$w = (1-\tau)(1-\alpha)y.$$ (5)

Government sector

The government budget is balanced in every period of time:

$$\tau y = g + z,$$ (6)

where $\tau y = \tau Ag^{1-\alpha} k^{\alpha}$ are total tax revenues and $g + z$ are total government spending, z being the expenditures allocated to improving the quality of the environment (abatement activities).

The allocation of total government spending between investment and abatement is made by following the rules:

$$g = (1-\phi)\,\tau Ag^{1-\alpha}k^{\alpha},$$ (7)

$$z = \phi\,\tau Ag^{1-\alpha}k^{\alpha},$$ (8)

where ϕ is the share of public spending devoted to abatement activities. We will refer to ϕ as public spending composition. From (7) and (8):

$$\frac{g}{k} = [(1-\phi)\,\tau A]^{1/\alpha},$$ (9)

$$z = \frac{\phi}{1-\phi}\,g.$$ (10)

Pollution function

We adopt the pollution function proposed by Smulders and Gradus (1996),[3] that is, pollution flow depends positively on capital stock use and negatively on abatement activities, i.e., $P = f(k, z)$, where $f_k > 0$ and $f_z < 0$.[4] Hence, by using (9) and (10) in this definition:

$$P = \frac{k^{z_1}}{z^{z_2}} = \frac{k^{z_1}}{\left[\dfrac{\phi}{1-\phi}g\right]^{z_2}} = \frac{k^{z_1 - z_2}}{\left[\phi^{\alpha}\left(1-\phi\right)^{(1-\alpha)}\tau A\right]^{z_2/\alpha}}. \tag{11}$$

Decentralized competitive equilibrium

Definition 1 *A competitive equilibrium for this economy is a set of allocations* $\{c, k, l\}_{t=0}^{\infty}$ *and a price system* $\{w, r\}_{t=0}^{\infty}$ *such that given a price system and a fiscal policy* $\{\tau, \phi\}_{t=0}^{\infty}$: *(i)* $\{c, k\}_{t=0}^{\infty}$ *maximizes household utility (1), subject to (2), and taking the pollution level* $\{P\}_{t=0}^{\infty}$ *and* $\{k_0\}$ *as given; (ii)* $\{k, l\}_{t=0}^{\infty}$ *satisfies the firm's profit maximization conditions, and (iii) all markets clear, that is,* $\{c, k, g, z\}_{t=0}^{\infty}$ *satisfies the aggregate resources constraint (12) below and the government budget constraint (6) and, in the labor market,* $l = 1$:

$$c + k = \left(1 - \tau\right) A\, g^{1-\alpha}\, k^{\alpha}. \tag{12}$$

By using (3), (4) and (9):

$$\frac{\dot{c}}{c} = \Psi\left(\tau, \phi\right) - \rho, \tag{13}$$

where:

$$\Psi\left(\tau, \phi\right) = \left(1 - \tau\right)\alpha A^{1/\alpha}\left[\tau\left(1 - \phi\right)\right]^{(1-\alpha)/\alpha}. \tag{14}$$

From (2), (4), (5) y (9):

$$\frac{\dot{k}}{k} = \frac{\Psi\left(\tau, \phi\right)}{\alpha} - \frac{c}{k}. \tag{15}$$

Note that, from (13), the long-run growth rate is positive if and only if $\Psi(\tau, \phi) > 0$. Hence, we will only consider balanced growth paths in which the time paths for τ, ϕ guarantee that this condition holds.

The decentralized equilibrium is globally and locally determinate (the proof is available upon request).

SECOND-BEST POLICY

The government chooses economic policy $\{\tau, \phi\}$ to maximize the utility of the representative household, subject to the competitive equilibrium conditions.

That is:

$$\underset{\{\tau,\phi\}}{Max} \quad \int_0^\infty e^{-\rho t} \left[\ln c - \eta \ln P \right] dt$$

subject to:

$$\dot{c} = c \left[\Psi(\tau,\phi) - \rho \right], \tag{16}$$

$$\dot{k} = \frac{\Psi(\tau,\phi)}{\alpha} k - c, \tag{17}$$

$$P = k^{\chi_1 - \chi_2} \left[\phi^\alpha \left(1-\phi\right)^{(1-\alpha)} \tau A \right]^{-\chi_2/\alpha}. \tag{18}$$

We refer to this optimal control problem as the *Ramsey problem* and to the associated allocations and policies as the *Ramsey allocations* and the *Ramsey plan*. We assume that the government has some power of commitment to bind itself to implement, at any future date, the plan it has announced in the first period.

The current-value Hamiltonian is:

$$H \equiv e^{-\rho t} \left[\ln c - \eta \left(\chi_1 - \chi_2 \right) \ln k + \eta \chi_2 \left(\ln \phi + \frac{1-\alpha}{\alpha} \ln\left(1-\phi\right) + \frac{1}{\alpha} \ln\left(\tau A\right) \right) \right] +$$

$$e^{-\rho t} \mu_1 \left[c \left(\Psi(\tau,\phi) - \rho \right) \right] + e^{-\rho t} \mu_2 \left[\frac{\Psi(\tau,\phi)}{\alpha} k - c \right]$$

where μ_1 and μ_2 are the dynamic multipliers associated to (16) and (17), respectively. The necessary conditions with respect to $\{c, k, \mu_1, \mu_2, \tau, \phi\}$ are given by:

$$-\frac{\partial H}{\partial c} = \dot{\mu}_1 - \rho\mu_1 \rightarrow \dot{\mu}_1 = -\frac{1}{c} + \mu_2 - \mu_1[\Psi - \rho] + \rho\mu_1, \tag{19}$$

$$-\frac{\partial H}{\partial k} = \dot{\mu}_2 - \rho\mu_2 \rightarrow \dot{\mu}_2 = \mu_2\left[\rho - \frac{\Psi}{\alpha}\right] + \eta(\chi_1 - \chi_2)\frac{1}{k}, \tag{20}$$

$$\frac{\partial H}{\partial \mu_1} = 0 \rightarrow \dot{c} = c[\Psi - \rho] \tag{21}$$

$$\frac{\partial H}{\partial \mu_2} = 0 \rightarrow \dot{k} = \frac{\Psi}{\alpha}k - c, \tag{22}$$

$$\frac{\partial H}{\partial \tau} = 0 \rightarrow 0 = \frac{\eta\chi_2}{\alpha}\frac{1}{\tau} + \mu_1 c\Psi_\tau + \mu_2\frac{k}{\alpha}\Psi_\tau, \tag{23}$$

$$\frac{\partial H}{\partial \phi} = 0 \rightarrow 0 = \eta\chi_2\frac{\alpha - \phi}{\alpha\phi(1-\phi)} + \mu_1 c\Psi_\phi + \mu_2\frac{k}{\alpha}\Psi_\phi$$
$$\rightarrow 0 = \eta\chi_2\frac{\alpha - \phi}{\phi} + \mu_1 c(\alpha - 1)\Psi + \mu_2\frac{(\alpha - 1)k}{\alpha}\Psi, \tag{24}$$

where $\Psi = \Psi(\tau, \phi)$, $\Psi_\tau = \frac{\partial \Psi}{\partial \tau} = \frac{1 - \tau - \alpha}{\tau\alpha}\Psi$, $\Psi_\phi = \frac{\partial \Psi}{\partial \phi} = \frac{\alpha - 1}{\alpha}\frac{\Psi}{(1-\phi)}$.

Proposition 1 *The optimal ratio of abatement to total government spending, ϕ, falls between 0 and α^2, and the optimal tax rate, τ, falls between $1-\alpha$ and 1, for all t.*

Proof. By using (23) and (24), we obtain that:

$$\phi = \alpha\left(1 - \frac{1-\alpha}{\tau}\right). \tag{25}$$

Hence, it is obvious that ϕ is monotonically increasing and concave in τ. Because ϕ and τ must be defined in the interval $(0,1)$, and given (25), it is readily obtained that the range for the function (25) is given by $[1-\alpha,1)$ and the domain is given by $[0,\alpha^2)$. ∎

The system (19)-(24) consists of six equations and six variables. We reduce its dimensionality to simplify computation. The new variables vector will be $\{x_1, x_2, \tau\}$, where x_1 and x_2 are two auxiliary variables defined as:

$$x_1 \equiv \frac{c}{k}, \tag{26}$$

$$x_2 \equiv \mu_2 c, \tag{27}$$

so that both x_1 and x_2 do not grow in the steady state. This implies that c and k grow at the same rate, denoted by γ.

Differentiating (26), (27) and (23) with respect to time and using (19) to (24), we obtain, after some algebra:

$$\dot{x}_1 = \left[\tilde{\Psi} \left(1 - \frac{1}{\alpha} \right) - \rho + x_1 \right] x_1, \tag{28}$$

$$\dot{x}_2 = \tilde{\Psi} \left(1 - \frac{1}{\alpha} \right) x_2 + \eta \left(\chi_1 - \chi_2 \right) x_1, \tag{29}$$

$$\dot{\tau} = \frac{1}{\eta \chi_2} \Theta \Omega, \tag{30}$$

where:[5]

$$\tilde{\Psi} = \tilde{\Psi}(\tau) = (1 - \tau) \alpha A^{1/\alpha} \left[(1 - \alpha)(\tau + \alpha) \right]^{(1-\alpha)/\alpha}, \tag{31}$$

$$\Theta = \Theta(\tau) = \frac{\alpha(1 - \tau)(\tau + \alpha)(1 - \tau - \alpha)^2 \tilde{\Psi}}{(1 - \tau - \alpha)\left[1 - \tau - \alpha - \alpha^2 \right] - \alpha(1 - \tau)(\tau + \alpha)}, \tag{32}$$

$$\Omega = \Omega(x_2, \tau) = \frac{\eta(\chi_1 - \chi_2)}{\alpha} - 1 - \frac{x_2(1 - \alpha)}{\alpha} - \frac{\rho \eta \chi_2}{(1 - \tau - \alpha)\tilde{\Psi}}. \tag{33}$$

Now, the steady state and the transitional dynamic properties of the Ramsey problem can be studied from the three-dimensional system (28)-(30) in $\{x_1, x_2, \tau\}$.

Steady state and global indeterminacy

This section focuses on the steady state or balanced growth path (BGP). An interior steady state is a vector $(\hat{x}_1, \hat{x}_2, \hat{\tau})$ satisfying the equations for the Ramsey problem (in particular, the equations (28)-(30)), such that if it is ever reached, the system will stay at that point forever $(\dot{x}_1 = \dot{x}_2 = \dot{\tau} = 0)$. Defining $\hat{\Psi} = \tilde{\Psi}(\hat{\tau})$, the steady state is given by:

$$\hat{x}_1 = \rho + \frac{(1-\alpha)\hat{\Psi}}{\alpha}, \tag{34}$$

$$\hat{x}_2 = \eta(\chi_1 - \chi_2)\left[1 + \frac{\rho\alpha}{(1-\alpha)\hat{\Psi}}\right] \tag{35}$$

and $\hat{\tau}$ is obtained from the implicit equation:

$$\hat{\Psi} = \frac{\rho\eta}{1-\eta(\chi_1-\chi_2)}\left[\chi_2 \frac{\hat{\tau}+\alpha}{(\hat{\tau}+\alpha)-1} - \chi_1\right]. \tag{36}$$

The first stage is to find the solution for (36) in the long-run tax rate, $\hat{\tau}$. Once it has been obtained, (34) can give the consumption to capital ratio in the steady state $(\hat{x}_1)$ and (35) determines the long-run value for the auxiliary variable $\hat{x}_2$.

Proposition 2 *A necessary condition for the long-run consumption growth rate to be positive is:*

$$\chi_2 \frac{\hat{\tau}+\alpha}{(\hat{\tau}+\alpha)-1} > \chi_1, \quad \text{together with} \quad 1 > \eta(\chi_1-\chi_2), \tag{37}$$

or

$$\chi_2 \frac{\hat{\tau}+\alpha}{(\hat{\tau}+\alpha)-1} < \chi_1, \quad \text{together with} \quad 1 < \eta(\chi_1-\chi_2), \tag{38}$$

Proof. From (21), the consumption growth rate is positive in equilibrium if and only if $\hat{\Psi} > \rho$. Hence, a necessary condition for positive growth is $\hat{\Psi} > 0$. As long as Proposition 1 implies that $\hat{\tau} \geqslant 1 - \alpha$, and hence, $\chi_2 (\hat{\tau} + \alpha) / [(\hat{\tau} + \alpha) - 1] > 0$, it is obvious from (36) that the condition for $\hat{\Psi} > 0$ is characterized by (37) or (38). ■

Proposition 3 *The second-best long-run consumption growth rate is positive if and only if* $\eta\chi_2 > \hat{\tau} + \alpha - 1$.

Proof. From (21), the consumption growth rate is positive in equilibrium if and only if $\hat{\Psi} > \rho$. From (36), $\hat{\Psi} > \rho$ if and only if:

$$\frac{\eta}{1-\eta(\chi_1-\chi_2)}\left[\chi_2\frac{\hat{\tau}+\alpha}{(\hat{\tau}+\alpha)-1}-\chi_1\right] > 1$$

After some algebra, the condition $\eta\chi_2 > \hat{\tau} + \alpha - 1$ can be obtained.∎

In order to study the conditions for global indeterminacy, first, we characterize the left-hand side and the right-hand side of (36), denoted, respectively, by:

$$LHS(\hat{\tau}) = \hat{\Psi} = (1-\hat{\tau})\alpha A^{\frac{1}{\alpha}}\left[(1-\alpha)(\hat{\tau}+\alpha)\right]^{\frac{1-\alpha}{\alpha}}$$

and:

$$RHS(\hat{\tau}) = \frac{\rho\eta}{1-\eta(\chi_1-\chi_2)}\left[\frac{\chi_2(\hat{\tau}+\alpha)}{(\hat{\tau}+\alpha)-1}-\chi_1\right].$$

Proposition 4 *The function LHS $(\hat{\tau})$ is monotonically decreasing and concave and its domain is given by* $[0, \alpha^2 A^{1/\alpha}(1-\alpha)^{1-\alpha/\alpha}]$, $\forall\, \hat{\tau} \in [1-\alpha,1)$

Proof. See Appendix. ∎

Proposition 5

$$\text{If } 1 < \eta\,(\chi_1\text{-}\chi_2) \text{ and } \chi_2\frac{1+\alpha}{\alpha} < (>)\,\chi_1$$

then the function RHS $(\hat{\tau})$ is monotonically increasing, concave, and its domain is $(-\infty,\kappa)$, *with* $\kappa > (<) 0$.

Proof. See Appendix. ∎

Proposition 6

$$\text{If } 1 > \eta\,(\chi_1\text{-}\chi_2) \text{ and } \chi_2\frac{1+\alpha}{\alpha} < (>)\,\chi_1$$

then the function RHS $(\hat{\tau})$ is monotonically decreasing and convex and its domain is (κ, ∞), *with* $\kappa < (>) 0$.

Proof. See Appendix. ∎

Let $\hat{\gamma}$ and $\hat{\phi}$ be the optimal long-run levels for the growth rate and the spending composition, respectively. Now, the conditions for global indeterminacy can be characterized. From the propositions above, four types of situations can be obtained:

I) If *RHS* $(\hat{\tau})$ is monotonically increasing, concave, and its domain is $(-\infty, \kappa)$, with $\kappa > 0$, then *RHS* $(\hat{\tau}) > 0$ for $\hat{\tau} \to \Gamma$. In this situation, there exists one single solution for $\hat{\tau}$, which verifies the necessary condition for positive growth rate.

II) If *RHS* $(\hat{\tau})$ is monotonically increasing, concave, and its domain is $(-\infty, \kappa)$, with $\kappa < 0$, then *RHS* $(\hat{\tau}) < 0$ for $\hat{\tau} \to \Gamma$. In this case, there is no solution.

III) If *RHS* $(\hat{\tau})$ is monotonically decreasing and convex, and its domain is (κ, ∞), with $\kappa < 0$, then *RHS* $(\hat{\tau}) < 0$ for $\hat{\tau} \to \Gamma$. Hence, there is one single solution for $\hat{\tau}$. In this case, the fulfilment of the necessary condition for positive long-run growth rate (Proposition 2) under the optimal tax rate must be checked, because it does not hold for $\hat{\tau} \to \Gamma$.

IV) If *RHS* $(\hat{\tau})$ is monotonically decreasing and convex, and its domain is (κ, ∞), with $\kappa > 0$, then *RHS* $(\hat{\tau}) > 0$, $\forall \hat{\tau} \in [1-\alpha, 1)$. Then there are three possibilities: First, *LHS* $(\hat{\tau})$ lies below *RHS* $(\hat{\tau})$, $\forall \hat{\tau} \in [1-\alpha, 1)$; in this case there is no solution. Second, *LHS* $(\hat{\tau})$ is tangent to *RHS* $(\hat{\tau})$; in this case, there is one single solution. Third, *LHS* $(\hat{\tau})$ lies above *RHS* $(\hat{\tau})$; in this case, there are two solutions. This last situation guarantees the existence of global indeterminacy; that is, there are two long-run optimal tax rates $\hat{\tau}_1$ and $\hat{\tau}_2$ such that $(1-\alpha) < \hat{\tau}_1, < \hat{\tau}_2 < 1$. Hence, there are two Ramsey allocations corresponding to two Ramsey plans. From (25) and (13), the optimal $(\hat{\phi}_1, \hat{\gamma}_1)$ and $(\hat{\phi}_2, \hat{\gamma}_2)$ can be obtained, verifying that $0 < \hat{\phi}_1 < \hat{\phi}_2$ α^2 and $\hat{\gamma}_1 > \hat{\gamma}_2$.[6]

Furthermore, if $\chi_1 > \chi_2$ the pollution growth rate will be positive and for $\hat{\tau}_1$ will be larger than for $\hat{\tau}_2$. On the other hand, the lower the tax rate is, the lower the share of public spending devoted to abatement is, and hence, the pollution level will also be larger for $\hat{\tau}_1$ *ceteris paribus*. If $\chi_1 > \chi_2$, the pollution growth rate is negative, and hence, pollution will fall at a faster rate the lower $\hat{\tau}$ is. However, the abatement activities are lower for $\hat{\tau}_1$ so that the pollution path for $\hat{\tau}_1$ lies above the pollution path for $\hat{\tau}_2$ only during the first periods.

The economic policy implication behind global indeterminacy is that two economies with the same fundamentals could choose different optimal policies (taxes and spending composition) that would yield different optimal paths of pollution. This result could explain the lack of convergence in the pollution levels of countries with similar fundamentals.

It is easy to provide numerical examples of a unique interior BGP or two interior BGPs. The first example leads to a unique interior BGP that can be obtained for the following set of structural parameters: $\alpha = 0.7$, $A = 1$, $\rho = $ -log (0.95), $\eta = 4$, $\chi_1 = 0.8$, $\chi_2 = 0.1$. The optimal tax rate corresponding to this framework is $\hat{\tau} = 0.88$, the optimal spending composition is $\hat{\phi} = 0.46$ and the optimal growth rate is $\hat{\gamma} = 0.009$. The second example yields a unique interior BGP. The set of structural parameters is: $\alpha = 0.7$, $A = 1$, $\rho = $ -log (0.95), $\eta = 3.311$, $\chi_1 = 0.4$, $\chi_2 = 0.2$. The optimal tax rate is $\hat{\tau} = 0.768$, the optimal spending composition is $\hat{\phi} = 0.427$ and the optimal growth rate is $\hat{\gamma} = 0.063$. Finally, the third example leads to two interior BGPs: The set of parameters is $\alpha = 0.7$, $A = 0.7$, $\rho = $ -log (0.95), $\eta = 1$, $\chi_1 = 0.7$, $\chi_2 = 0.6$. For the first BGP, the optimal tax rate is $\hat{\tau}_1 = 0.58$, the optimal spending composition is $\hat{\phi}_1 = 0.34$ and the optimal growth rate is $\hat{\gamma}_1 = 0.067$. The second BGP is defined by $\hat{\tau}_2 = 0.78$, $\hat{\phi}_2 = 0.43$, $\hat{\gamma}_2 = 0.0145$.

In our model, the interaction between the two externalities (pollution in utility and public productive services) is key for obtaining the indeterminacy result. If only one externality is considered, the model does not exhibit indeterminacy. Proof is available upon request. A similar proof can be found in Park and Philippopoulos (2004).

Transitional dynamics and local indeterminacy

Linearizing the dynamic system (28)-(30) around the BGP in (34)-(36) allows us to obtain the following linear system:

$$
\begin{bmatrix} \dot{x}_1 \\ \dot{x}_2 \\ \dot{\tau} \end{bmatrix} = \begin{bmatrix} \hat{x}_1 & 0 & \hat{x}_1 \hat{\Psi}_\tau \left(1 - \dfrac{1}{\alpha}\right) \\ \eta(\chi_1 - \chi_2) & \hat{\Psi}\left(1 - \dfrac{1}{\alpha}\right) & \hat{x}_2 \hat{\Psi}_\tau \left(1 - \dfrac{1}{\alpha}\right) \\ 0 & \hat{\Xi} & \rho \end{bmatrix} \begin{bmatrix} x_1 - \hat{x}_1 \\ x_2 - \hat{x}_2 \\ \tau - \hat{\tau} \end{bmatrix}, \quad (39)
$$

where:

$$
\hat{\Psi}_\tau = \frac{\partial \tilde{\Psi}(\hat{\tau})}{\partial \tau} = \frac{1 - \hat{\tau} - \alpha - \alpha^2}{\alpha(1 - \hat{\tau})(\hat{\tau} + \alpha)} \hat{\Psi},
$$

$$
\hat{\Xi} = -\frac{\hat{\Psi}_\tau \alpha (1 - \alpha)(1 - \hat{\tau})(\hat{\tau} + \alpha)(1 - \hat{\tau} - \alpha)^2}{\alpha \eta \chi_2 \left[(1 - \hat{\tau} - \alpha)(1 - \hat{\tau} - \alpha - \alpha^2) - \alpha(1 - \hat{\tau})(\hat{\tau} + \alpha)\right]}.
$$

Let Γ be the Jacobian matrix in (39) and the characteristic equation of Γ is:

$$w^3 - tr\left(\Gamma\right)w^2 + \Omega\left(\Gamma\right)w - \det\left(\Gamma\right) = 0, \text{ where: } tr\left(\Gamma\right) = 2\rho, \text{ and:}$$

$$\det\left(\Gamma\right) = \left(1 - \frac{1}{\alpha}\right)\rho\hat{x}_1\hat{\Psi}\left[1 + \frac{\left(\chi_1 - \chi_2\right)\left(1 - \hat{\tau} - \alpha\right)^2 \hat{\Psi}_r}{\chi_2\left[\left(1 - \hat{\tau} - \alpha\right)\hat{\Psi}_r - \hat{\Psi}\right]}\right],$$

$$\Omega\left(\Gamma\right) = \left(1 - \frac{1}{\alpha}\right)\left[\hat{x}_2\hat{\Psi}_r\hat{\Xi} + \hat{x}_1\hat{\Psi} + \rho\hat{\Psi}\right] + \hat{x}_1\rho.$$

We will use Routh's Theorem in our analysis (see Gantmacher (1960, Chapter XV)). In our particular case, this theorem says that the number of roots with positive real parts is equal to the number of variations of sign in the sequence: -1, tr (Γ), $-\Omega$ (Γ) + det (Γ) / tr (Γ), det (Γ). From now on, this sequence will be called Routh's Sequence.

Although it is not possible from an analytical point of view to know the sign for the third term in the sequence, we can claim the following sufficient condition for local indeterminacy.

Proposition 7

If $\displaystyle 1 + \frac{\left(\chi_1 - \chi_2\right)\left(1 - \hat{\tau} - \alpha\right)^2 \hat{\Psi}_r}{\chi_2\left[\left(1 - \hat{\tau} - \alpha\right)\hat{\Psi}_r - \hat{\Psi}\right]} > 0$ *then there is local indeterminacy.*

Proof.

$$\text{If } 1 + \frac{\left(\chi_1 - \chi_2\right)\left(1 - \hat{\tau} - \alpha\right)^2 \hat{\Psi}_r}{\chi_2\left[\left(1 - \hat{\tau} - \alpha\right)\hat{\Psi}_r - \hat{\Psi}\right]} > 0 \text{ then det } (\Gamma).$$

Hence, given Routh's Sequence, there are only two variations of sign, whatever the sign of $-\Omega$ (Γ) + det (Γ) / tr (Γ) is. Consequently, there are two roots with a positive real part and one root with negative real part. Since the three variables in the system (39) are jump, there is local determinacy only if the three roots have positive real part. Hence, under the condition:

$$1 + \frac{\left(\chi_1 - \chi_2\right)\left(1 - \hat{\tau} - \alpha\right)^2 \hat{\Psi}_r}{\chi_2\left[\left(1 - \hat{\tau} - \alpha\right)\hat{\Psi}_r - \hat{\Psi}\right]} > 0,$$

the BGP is locally indeterminate.

Note that indeterminacy could also arise if:

$$1+\frac{(\chi_1-\chi_2)(1-\hat{\tau}-\alpha)^2\hat{\Psi}_r}{\chi_2[(1-\hat{\tau}-\alpha)\hat{\Psi}_r-\hat{\Psi}]}<0 \text{ and } -\Omega(\Gamma)+\det(\Gamma)/tr(\Gamma)>0.$$

In this case, there are two roots with negative real part because only one variation of sign would exist in Routh's Sequence.

The implication of local indeterminacy is that there will be multiple paths converging to a given steady state. In our model, indeterminacy is able to explain why two countries with similar fundamentals (preferences, technology and initial capital stock) might display different pollution paths even if such economies choose the same public policies.

In this sense, the result of local indeterminacy is able to explain the lack of empirical regularity in the pollution experiences of different economies; for example, the difficulty to find regularity with respect to the Kuznets curve, the results being dependent on the sample of countries and years studied. That is, the past pollution experience of a developed country does not guarantee that a developing country will follow the same path over time even if the initial conditions and the environmental policies are similar.

The numerical examples shown in the preceding subsection are also useful for illustrating different situations of local indeterminacy.

The first example, defined by the following parameter set, $\alpha=0.7$, $A=1$, $\rho=-\log(0.95)$, $\eta=4$, $\chi_1=0.8$, $\chi_2=0.1$, displays local and global determinacy. That is, there is one single converging path towards the single steady state. The second example ($\alpha=0.7$, $A=1$, $\rho=-\log(0.95)$, $\eta=3.311$, $\chi_1=0.4$, $\chi_2=0.2$) also displays local and global determinacy. The two BGPs of the third example ($\alpha=0.7$, $A=0.7$, $\rho=-\log(0.95)$, $\eta=1$, $\chi_1=0.7$, $\chi_2=0.6$) are locally indeterminate. That is, there is a continuum of equilibria converging to each BGP.

It is also possible to find examples of global indeterminacy with one locally determinate BGP and one locally indeterminate BGP. The following set of parameters $\alpha=0.7$, $A=1$, $\rho=-\log(0.95)$, $\eta=3.310$, $\chi_1=0.4$, $\chi_2=0.2$, yields two BGPs characterized by $\hat{\tau}_1=0.76$, $\hat{\phi}_1=0.42$, $\hat{\gamma}_1=0.067$ and $\hat{\tau}_2=0.78$, $\hat{\phi}_2=0.43$, $\hat{\gamma}_2=0.059$. The first BGP is locally determinate and the second locally indeterminate.

CONCLUSION

In a second-best framework, we study the dynamic properties of a general equilibrium model of endogenous growth with environmental externalities and public abatement activities. In particular, we investigate the conditions for global and local indeterminacy, and their implications for the governmental control of pollution.

We show that when we include pollution in the utility function together with public abatement activities and public productive services, the introduction of endogenously chosen economic policy can generate global and local indeterminacy; that is, there can be two steady states (global indeterminacy), both of which can be locally indeterminate (that is, there are multiple converging paths to each state). If only one externality is considered, the model does not exhibit indeterminacy.

Global and local indeterminacy are able to explain why two countries with similar fundamentals (preferences, technology and initial capital stock) might display different short and long-run pollution paths regardless of the policies implemented.

APPENDIX

Proof of Proposition 4

Given $LHS(\hat{\tau}) = \hat{\Psi} = (1 - \hat{\tau})\alpha A^{1/\alpha}[(1 - \alpha)(\hat{\tau}+\alpha)]^{(1-\alpha)/\alpha}$, first we show that it is decreasing for $\hat{\tau} \in [1-\alpha,1)$.

Since: $\partial LHS(\hat{\tau})/\partial\hat{\tau} = A^{1/\alpha}(1 - \alpha)^{(1-\alpha)/\alpha}(\hat{\tau}+\alpha)^{(1-2\alpha)/\alpha}[1-\alpha - \alpha^2 - \hat{\tau}]$, then: $\partial LHS(\hat{\tau})/\partial\hat{\tau} < 0$, for $\hat{\tau} > 1 - \alpha - \alpha^2$. Since $\hat{\tau} > 1 - \alpha > 1 - \alpha - \alpha^2$, then $LHS(\hat{\tau})$ is decreasing for the range of $\hat{\tau}$.

Second, we show that $LHS(\hat{\tau})$ is concave in $\hat{\tau}$. $\partial^2 LHS(\hat{\tau})/\partial\hat{\tau}^2 = \hat{\Psi}[\alpha(1-\hat{\tau})(\hat{\tau}+\alpha)]^{-2}[\alpha(1-\alpha-\alpha^2-\hat{\tau})(2\hat{\tau}+\alpha)-\alpha(1-\hat{\tau})(\hat{\tau}+\alpha] < 0$ since $\hat{\tau} > 1-\alpha-\alpha^2$, as we have shown previously.

Furthermore: $\lim_{\hat{\tau}\to(1-\alpha)^+}\hat{\Psi} = \alpha^2 A^{1/\alpha}[1-\alpha]^{(1-\alpha)/\alpha} > 0; \lim_{\hat{\tau}\to1^-}\hat{\Psi} = 0$

Therefore, the function is positive $\forall \hat{\tau} \in [1-\alpha,1)$.

Proof of Propositions 5 and 6

Define $\Delta = 1-\eta(\chi_1 - \chi_2)$ and $\Pi = \chi_2\dfrac{1+\alpha}{\alpha} - \chi_1$.

When $\hat{\tau} > (1-\alpha)$, $RHS(\hat{\tau}) = \dfrac{\rho\eta}{\Delta}\left[\chi_2\dfrac{\hat{\tau}+\alpha}{\hat{\tau}+\alpha-1} - \chi_1\right]$

is continuous, and decreasing (increasing) in $\hat{\tau}$ if $\Delta > (<) 0$:

$$\frac{\partial RHS(\hat{\tau})}{\partial \hat{\tau}} = -\frac{\rho\eta}{\Delta}\frac{1}{\left[(\hat{\tau}+\alpha)-1\right]^2}, \frac{\partial^2 RHS(\hat{\tau})}{\partial \hat{\tau}^2} = \frac{2\rho\eta}{\Delta}\frac{1}{\left[(\hat{\tau}+\alpha)-1\right]^3}, \text{ and hence:}$$

I) If $\Delta < 0$ and $\Pi < 0$:

$$\frac{\partial RHS(\hat{\tau})}{\partial \hat{\tau}} > 0; \frac{\partial^2 RHS(\hat{\tau})}{\partial \hat{\tau}^2} < 0; \lim_{\hat{\tau}\to(1-\alpha)^+} RHS(\hat{\tau}) = -\infty; \lim_{\hat{\tau}\to 1^-} RHS(\hat{\tau}) = \kappa > 0,$$

and therefore, $RHS(\hat{\tau})$ is increasing and concave, with domain $(-\infty, \kappa)$.

II) If $\Delta < 0$ and $\Pi > 0$:

$$\frac{\partial RHS(\hat{\tau})}{\partial \hat{\tau}} > 0; \frac{\partial^2 RHS(\hat{\tau})}{\partial \hat{\tau}^2} < 0; \lim_{\hat{\tau}\to(1-\alpha)^+} RHS(\hat{\tau}) = -\infty; \lim_{\hat{\tau}\to 1^-} RHS(\hat{\tau}) = \kappa < 0,$$

and therefore, $RHS(\hat{\tau})$ is increasing and concave, with domain $(-\infty, \kappa)$.

III) If $\Delta > 0$ and $\Pi < 0$

$$\frac{\partial RHS(\hat{\tau})}{\partial \hat{\tau}} < 0; \frac{\partial^2 RHS(\hat{\tau})}{\partial \hat{\tau}^2} > 0; \lim_{\hat{\tau}\to(1-\alpha)^+} RHS(\hat{\tau}) = \infty; \lim_{\hat{\tau}\to 1^-} RHS(\hat{\tau}) = \kappa < 0,$$

and therefore, $RHS(\hat{\tau})$ is decreasing and convex, with domain (κ, ∞).

IV) If $\Delta > 0$ and $\Pi < 0$

$$\frac{\partial RHS(\hat{\tau})}{\partial \hat{\tau}} < 0; \frac{\partial^2 RHS(\hat{\tau})}{\partial \hat{\tau}^2} > 0; \lim_{\hat{\tau}\to(1-\alpha)^+} RHS(\hat{\tau}) = \infty; \lim_{\hat{\tau}\to 1^-} RHS(\hat{\tau}) = \kappa > 0,$$

and therefore, $RHS(\hat{\tau})$ is decreasing and convex, with domain (κ, ∞).

NOTES

* The authors would like to thank professors A. Novales, S. Smulders and the participants at the First Atlantic Workshop on Energy and Environmental Economics for their helpful comments. The authors are also grateful for financial support from Fundación Ramón Areces, from the Spanish Ministry of Education through grant BEC2003-03965 and from Fundación CentrA.

1. Under a stochastic framework, local indeterminacy guarantees existence of a continuum of

sunspot stationary equilibria, i.e., stochastic rational expectations equilibria determined by perturbations unrelated to the uncertainty in economic fundamentals.

2. They include environmental quality as a positive externality in the utility function. However, environmental quality is inversely related to pollution, leading to the same welfare function employed in our model.

3. Mohtadi (1996) adopts a similar function for the level of the environmental quality, where the environmental quality is inversely related to pollution flow.

4. We define $f_x = \partial f / \partial x$, for $x = k, z$.

5. Note that by using (25) in (14), we obtain (31).

6. Note that in this case $\hat{r}$ verifies the necessary condition for a positive growth rate (Proposition 2).

REFERENCES

Aiyagari, S.R. (1995), 'The econometrics of indeterminacy: an applied study, a coment', *Carnegie-Rochester Conf. Ser. Public Policy*, **43**, 273-282.

Benhabib, J. and R.E.A. Farmer (1996), 'Indeterminacy and sector specific externalities', *Journal of Monetary Economics*, **37**, 421-443.

Benhabib, J. and R.E.A. Farmer (1994), 'Indeterminacy and increasing returns', *Journal of Economic Theory*, **63**, 19-41.

Benhabib, J. and K. Nishimura (1998), 'Indeterminacy and sunspots with constant returns', *Journal of Economic Theory*, **81**, 58-96.

Benhabib, J. and R. Perli (1994), 'Uniqueness and indeterminacy: transitioal dynamics with multiple equilibria', *Journal of Economic Theory*, **63**, 113-142.

Bennett, R.L. and R.E.A. Farmer (2000), 'Indeterminacy with non-separable utility', *Journal of Economic Theory*, **93**, 118-143.

Bovenberg, A.L. and R.A. de Mooij (1997), 'Environmental tax reform and endogenous growth', *Journal of Public Economics*, **63**, 207-237.

De Bruyn, S.M. and R.J. Heintz (1999), 'The environmental Kuznets curve hypothesis', in J. van den Bergh, *Handbook of Environmental and Resource Economics*, Cheltenham: Edward Elgar.

Farmer, R.E.A. and J.T. Guo (1994), 'Real business cycles and the animal spirits hypothesis', *Journal of Economic Theory*, **63**, 42-73.

Fernández, E., A. Novales and J. Ruiz (2003), 'Indeterminacy under non-separability of public consumption and leisure in the utility function', *Economic Modelling*, **21**, 409-428.

Gantmacher, F.R. (1960), *The Theory of Matrices*, vol. II, New York, Chelsea.

Ligthart, J.E. and F. van der Ploeg (1994), 'Pollution, the cost of public funds and endogenous growth', *Economics Letters*, **46**, 351-361.

Mohtadi, H. (1996), 'Environment, growth, and optimal policy design', *Journal of Public Economics*, **63**, 119-140.

Park, H. and A. Philippopoulos (2004), 'Indeterminacy and fiscal policies in a growing economy', *Journal of Economic Dynamics and Control*, **28**, 645-660.

Perli, R. (1998), 'Indeterminacy, home production and the business cycle', *Journal of Monetary Economics*, **41**, 105-125.

Smulders, S. and R. Gradus (1996), 'Pollution abatement and long-term growth', *European Journal of Political Economy*, **12**, 505-532.

Weder, M. (1998), 'Fickle consumers, durable goods and business cycles', *Journal of Economic Theory*, **81**, 37-57.

Wen, Y. (1998), 'Capacity utilization under increasing returns to scale', *Journal of Economic Theory*, **81**, 7-36.

Xie, D. (1994), 'Divergence in economic performance: transitional dynamics with multiple equilibria', *Journal of Economic Theory*, **63**, 97-112.

11. Energy-saving technological progress in a vintage capital model

Agustín Pérez-Barahona and Benteng Zou

INTRODUCTION

Fossil fuel is an essential input throughout all modern economies. However, the reduced availability of this basic production input, and the stabilization of greenhouse gases concentration – which requires reductions in fossil fuel energy use – have a negative impact on GDP and economic growth through cutbacks in energy use (Dasgupta and Heal, 1974; Hartwick and Olewiler, 1986; Smulders and Nooij, 2003).

Nevertheless, some authors (Newell *et al.*, 1999; Boucekkine and Pommeret, 2004; Carraro *et al.*, 2003) suggest that this trade-off between energy reduction and growth could be less severe if energy conservation were increased by energy-saving technologies. This chapter studies the former hypothesis and, in particular, the effect of a tax on the energy expenditure of firms as a way to encourage investments in energy-saving technologies.

Our model is an extension of Boucekkine and Pommeret's (2004) contribution to the general equilibrium case. The general equilibrium framework is very interesting because it allows us to study the global effect of environmental policies on the economy, and its relation with the scarce energy supply and the spreading out of energy-saving technologies. Here, we consider (exogenous) technological progress to be embodied in new capital goods, which are introduced in the economy through vintage technology with endogenous obsolescence (scrapping) rule. Considering the general equilibrium case without linearity, we assume constant scrapping age in the long run (Terborgh, 1949 and Smith, 1961 result).[1]

This chapter is organized as follows. In the next section, we describe the model, consumer behaviour and the rules that depict both the optimal investment and the scrapping behaviour of the firms. The balanced growth path (BGP) is presented in the third section, where we show the necessary conditions for its existence. Section four develops the static comparative analysis of our endogenous variables, along the BGP defined above. Finally, we make some concluding remarks.

THE MODEL

Let us consider an economy with a constant population level, where the labour market is perfectly competitive. The production sector produces only one final good (numeraire), which can be assigned to consumption or investment. One important difference with respect to the model of Boucekkine *et al.* (1997) is that, in our case, inputs are produced by means of non-linear technology. Moreover, this technology is defined by vintage capital. The input market is supposed to be monopolistically competitive to allow for a concave profit function in the inputs sector. Furthermore, we assume the available energy supply is exogenous.

Household

The household solves the following standard inter-temporal maximization problem with constant relative risk aversion (CRRA) instantaneous utility function:

$$\max_{c(t)} \int_0^\infty \frac{c(t)^{1-\theta}}{1-\theta} \exp\{-\rho t\} dt, \tag{1}$$

subject to the budget constraint and the corresponding transversality condition:

$$\dot{a}(t) = r(t)a(t) - c(t),$$

$$a(0) \text{ given}, \tag{2}$$

$$\lim_{t \to \infty} a(t) \exp\left\{ -\int_0^t r(z)dz \right\} = 0$$

where $c(t)$ is *per-capita* consumption, $a(t)$ is *per-capita* assets held by the household and the interest rate $r(t)$ is taken as given by the household. $\theta (> 0)$ measures the constant relative risk aversion, and $\rho (> 0)$ is the time preference parameter. It is easy to check that along the optimal path of consumption, the consumption follows:

$$r(t) = \rho + \theta \frac{\dot{c}(t)}{c(t)} \tag{3}$$

Final good firm

The final good is produced competitively by a representative firm solving the following optimal profit problem:

$$\max_{y_j(t)} y(t) - \int_0^1 p_j(t) y_j(t) dj$$

where the *per-capita* production $y(t)$ is given by a constant elasticity of substitution (CES) production technology:

$$y(t) = \left(\int_0^1 y_j(t)^{\frac{\varepsilon-1}{\varepsilon}} dj \right)^{\frac{\varepsilon}{\varepsilon-1}}$$

which is defined over a continuum of inputs $y_j(t)$ with $j \in [0,1]$. Moreover, constant elasticity of substitution $\varepsilon > 1$ is assumed. Furthermore, prices are given by:

$$p_j(t) = \left(\frac{y_j(t)}{y(t)} \right)^{-\frac{1}{\varepsilon}}$$

from the standard monopolistic competitive economy (Dixit and Stiglitz, 1977), and they are taken as given by the final good firm.

Input firm

Producing in a monopolistically competitive market, the representative input-j firm maximizes its profits:

$$\max_{y_j(t), i_j(t), T_j(t), p_j(t)} \int_0^\infty \left[p_j(t) y_j(t) - i_j(t) - e_j(t) P_e(1+Z) \right] \exp\left\{ -\int_0^t r(z) dz \right\} dt \tag{4}$$

where:

$$y_j(t) = A \left(\int_{t-T_j(t)}^t i_j(z)dz \right)^\alpha, \quad 0 < \alpha < 1 \tag{5}$$

$$e_j(t) = \int_{t-T_j(t)}^t i_j(z) \exp\{-\gamma z\}dz, \quad 0 < \gamma < 1 \tag{6}$$

$$p_j(t) = \left(\frac{y_j(t)}{y(t)} \right)^{-\frac{1}{\varepsilon}} \tag{7}$$

with the initial conditions $i_j(t)$ given for all $t \le 0$, $\forall j \in [0,1]$.

$e_j(t)$ and $P_e(t)$ are, respectively, the demand and the price of energy, which are endogenous. Z is the energy expenditure tax defined by the government.[2] $i_j(t)$ is the investment of the representative input-j firm. The output and the price for input j are respectively represented by $y_j(t)$ and $p_j(t)$. The price of input j and the final good production *per-capita*, $y(t)$, are taken as given by the monopoly. Equation (5) is our non-linear technology defined over vintage capital. The energy demand is obtained by equation (6). Here $\gamma > 0$ represents the rate of energy-saving technological progress and $T_j(t)$ is the age of the oldest operating machines or scrapping age. Considering monopolistic competition, the inverse demand function is given by equation (7).

Notice that the new technology is more energy-saving. Moreover, it is important to observe that we assume complementarity between capital and energy (Leontieff technology). Indeed, each vintage t has an energy requirement $i_j(t) \exp\{-\gamma t\}$. This assumption is unfailing from numerous studies; for instance Hudson and Jorgenson (1975), and Berndt and Wood (1974).

We define the optimal life of machines of vintage t as:

$$J_j(t) = T_j(t + J_j(t)) \tag{8}$$

Notice that $T_j(t) = J_j(t - T_j(t))$

Substituting (5)–(8) into (4), and considering the symmetric case, we can simplify our problem as follows:

$$\max_{i(t), T(t)} \int_t^{t+J(t)} \left[A \left(\int_{\tau-T(\tau)}^\tau i(z)dz \right)^\alpha - i(\tau) - P_e(\tau)(1 + Z) \int_{\tau-T(\tau)}^\tau i(z) \exp\{-\gamma z\}dz \right] \tag{9}$$

$$\exp\left\{ -\int_t^\tau r(z)dz \right\} d\tau$$

where $r(t)$ is given by (3).

From the first-order condition (FOC) for $i(t)$, we obtain the *optimal investment rule*:

$$\int_t^{t+J(t)} \alpha A \left(\int_{\tau-T(\tau)}^{\tau} i(z)dz \right)^{\alpha-1} \exp\left\{ -\int_t^{\tau} r(z)dz \right\} d\tau =$$
$$1 + \int_t^{t+J(t)} (1+Z)P_e(\tau)\exp\{-\gamma t\}\exp\left\{ -\int_t^{\tau} r(z)dz \right\} d\tau \qquad (10)$$

where the left hand side (LHS) is the discounted marginal productivity during the whole lifetime of the capital acquired in t; 1 is the marginal purchase cost at t, normalized to one; and the second term on the right-hand side (RHS) is the discounted operation cost at t. The *optimal investment rule* establishes that firms should invest at time t until the discounted marginal productivity during the whole lifetime of the capital acquired in t exactly compensates for both their discounted operation cost and their marginal purchase cost at t.

The FOC for $T(t)$ leads to the *optimal scrapping rule*:

$$A\alpha \left(\int_{t-T(t)}^{t} i(z)dz \right)^{\alpha-1} = P_e(t)(1+Z)\exp\{-\gamma(t-T(t))\} \qquad (11)$$

The *optimal scrapping rule* states that a machine should be scrapped as soon as its marginal productivity (which is the same for any machine whatever its age) no longer covers its operation cost (which rises with its age).

To summarize, the (decentralized) equilibrium of our economy is characterized by the equations (2)–(3) (household side), the equations (5)–(8), the *optimal investment rule*, the *optimal scrapping rule*, and the following three additional equations to close the model: $y(t) = c(t) + i(t)$, $i(t) = \dot{a}(t)$, and $e(t) = e_s(t)$, the equilibrium condition in the energy market. $e_s(t)$ is the available energy supply,[3] which is assumed to be exogenous in our model.

BALANCED GROWTH PATH

Let us define our balanced growth path (BGP) equilibrium as the situation where all the endogenous variables grow at a constant rate, with constant and finite scrapping age $T(t) = J(t) = T$ (Terborgh-Smith result).[4] Pérez-Barahona and Zou (2004) explicitly show that, with decreasing return to scale $(0 < \alpha < 1)$, there is no growth for output, investment and consumption in the long run. Moreover, they observe that, since energy market equilibrium requires energy demand to equal the energy supply $e_s(t)$, BGP (for the case of decreasing returns to scale) is only possible under a gloomy scenario (*i.e.*, $e_s(t) = \bar{e}_s \exp\{-\gamma t\}$). In the following, we are going to deduce the equilibrium conditions which will be used in the next section.

Obviously, at equilibrium, the interest rate is constant $r(t) = \rho$.

Differentiating (10) and rearranging terms, from (11) we obtain:

$$\exp\{\gamma T\} - 1 - \frac{\gamma}{\gamma_{P_e} - \gamma}\left(\exp\{(\gamma_{P_e} - \gamma)J\} - 1\right) = \frac{\gamma}{(1 + Z)\overline{P}_e}\exp\{(\gamma - \gamma_{P_e})t\}$$

where γ_{P_e} and $\overline{P}_e$ are, respectively, the growth rate and the level of the energy prices. The LHS is constant for any t in the equilibrium, and the RHS is a function of t. So the equality holds if and only if $\gamma = \gamma_{P_e}$. As in the standard growth model, this result states that, in terms of energy saving, energy prices grow at the same rate as productivity.

Defining the capital stock as:

$$K(t) = \int_{t-T(t)}^{t} i(z)dz$$

we have $\overline{K} = i^*\overline{T}$ along the BGP, where i^* and $\overline{K}$ are, respectively, the equilibrium investment and the optimal scrapping age. Then, since $\gamma = \gamma_{P_e}$, we have:

$$A\alpha\overline{K}^{\alpha-1} = (1 + Z)\overline{P}_e \exp\{\gamma\overline{T}\} \tag{12}$$

The economic intuition for this absence of long-run investment growth is the following. On the one hand, because of the assumption of decreasing returns to scale and, on the other hand, because here we find that both the scrapping age and exogenous energy-saving technical progress are not strong enough to overcome these decreasing returns. The reason is that our framework considers a CRRA instantaneous utility function, and as a consequence, the interest rate is constant in the long run. Then, consistent with the Terborgh-Smith result, the scrapping age is also constant along the BGP. Taking the optimal investment rule in the long run:

$$\frac{A\alpha(i^*\overline{T})^{\alpha-1}}{\rho}(1 - \exp\{-\rho\overline{T}\}) = 1 + \frac{(1 + Z)\overline{P}_e}{\rho - \gamma}(1 - \exp\{-(\rho - \gamma)\overline{T}\}) \tag{13}$$

It is straightforward to show that the discounted operation cost is constant because the effect of the energy-saving technical progress (γ) is offset by the decreasing available energy supply. Hence, as the marginal purchase cost ($=1$) remains constant, the investment has to be constant along the BGP. The economic interpretation of this result is the following. First of all, the scrapping age is assumed to be constant and finite, following the Terborgh-Smith result. Second, the energy supply has two effects on the economy: one through energy prices; and the other through the Leontieff production function, which models the idea of minimum energy requirements for using a machine. Considering a gloomy scenario, energy prices increase because of the decreasing available energy supply. Now, the rise in energy prices offsets the effect of the energy-saving technical progress. Moreover, the decreasing available energy supply is going to affect growth negatively through the minimum energy requirement (Leontieff technology). Hence, since we consider long-run replacement constant, there is no long-run growth.

To finish this section, we point out that our findings are not standard results. Indeed, the neoclassical models would achieve exogenous growth. For example, the models of Solow-Swan and Ramsey, with exogenous technical progress, describe economies which grow at the growth rate of both population and exogenous technical progress. However, we find here that the reduced availability of energy (non-renewable resource) offsets the (exogenous) energy-saving technical progress. As a consequence, since we consider equilibrium with constant scrapping age, our economy does not achieve long-run growth. Furthermore, considering a canonical vintage capital model with Arrowian learning-by-doing technical progress, d'Autume and Michel (1993) find that decreasing returns to scale 'kill' growth in the long run. Our model is mathematically close to their economy, taking energy instead of labour. Finally, we have to observe that our result is consistent with the partial equilibrium model of Boucekkine and Pommeret (2004), which also depicts no growth along the BGP.

STATIC COMPARATIVE OF ENERGY EXPENDITURE TAX

We consider the effect of modifications in the energy expenditure tax on the endogenous variables along the BGP. In our model, this effect is mainly explained by the behaviour of the scrapping age and the investment. The behaviour of the optimal investment rule is given by (13), and the optimal scrapping age is described by:

$$A\alpha(i^*\overline{T})^{\alpha-1} = (1+Z)\overline{P}_e \exp\{\gamma\overline{T}\} \qquad (14)$$

Since along the BGP the *optimal scrapping rule* (13) and the *optimal investment rule* (14) are functions of i^*, $\overline{T}$ and $\overline{P}_e$, we need one more equation to describe completely the behaviour of both the scrapping age and the investment. We find the third equation from the equilibrium condition of the energy market.

The energy demand is given by the equation:

$$e(t) = \int_{t-\bar{T}}^{t} i \exp\{-\gamma z\} dz = \frac{i}{\gamma} \exp\{-\gamma t\}(\exp\{\gamma \bar{T}\} - 1) = e_s(t) \qquad (15)$$

Since we have assumed an exogenous long-run energy supply $e_s(t)^* = \bar{e}_s \exp\{-\gamma t\}$ to have BGP, then we get:

$$\bar{e}_s = \frac{\overset{*}{i}}{\gamma}(\exp\{\gamma \bar{T}\} - 1) \qquad (16)$$

by equalizing the energy demand and the available energy supply. Therefore, the values of optimal investment and optimal scrapping age and the level of energy prices along the BGP are given by the equations (13), (14) and (16), which form a static (simultaneous) system of non-linear equations, taken the values of the parameters as given. Moreover, solving this system for different parameter values, we can analyze the static comparative of our model. In the Appendix, we include the parameterization and results of our static comparative exercise (see Table 11.1 and Table 11.2).

Our static comparative exercise analyzes the effect of an increase in the energy tax rate on our economy. In addition, considering such a static comparative analysis, we can describe some of the differences between economies with different energy tax rates. The purpose of this section is to describe the effects of an increase of Z on the optimal scrapping age ($\bar{T}$), the optimal investment (i^*) and the output (y^*).

A first approach is through a *pure analytical method*. From equation (16) we can obtain an expression for investment as a function of the scrapping age. Applying this to equation (14) (*optimal scrapping rule* in the long run) and differentiating that expression with respect to Z, we get a function $F_1(\partial \bar{T} / \partial Z; \partial \bar{P}_e / \partial Z)$. To obtain the value of $\partial \bar{T} / \partial Z$ and $\partial \bar{P}_e / \partial Z$, we need a second equation $F_2(\partial \bar{T} / \partial Z; \partial \bar{P}_e / \partial Z)$. This expression comes from the differentiation of the *optimal investment rule* in the long run (16), after considerable manipulations. Although this method allows us to obtain the analytical value of the derivatives, we cannot determine the sign of these for general values of the parameters, not even by imposing restrictions on some of them.

So, let us consider an *alternative method*. This procedure is a combination of the analytical approach and the numerical solution of the static system of non-linear equations given by the expressions (13), (14) and (16). By numerical methods – taking the empirical values of the parameters – we can solve that system of non-linear equations for different values of the energy expenditure tax $Z \in (0,1)$. So, we can determine the sign of the derivatives simply by plotting $\bar{T}$, i^* and $\bar{P}_e$ against Z. In particular, here we study the case of a high tax on the energy expenditure of firms. Bailey (2002) observed that taxes in the UK comprised 81.5 percent of total fuel prices. Following such an example, and consistently with the aim of the Kyoto Protocol, we assume a $Z = 0.80$. The results of the simulation suggest that an increase in the energy expenditure tax boosts the optimal replacement age, and decreases both the optimal investment and the level of energy prices. The explanation of this inverse relation between the level of energy prices and the energy expenditure tax comes directly from the assumption of an exogenous long-run energy supply. Considering the economy in the long run, if we increase that tax, the available energy supply is not affected (notice that the economy is in steady state) because it is exogenous and always decreasing in time. However, for a fixed level of production and scrapping age, the energy demand is reduced since energy is now more expensive. As a result, the level of energy prices decreases.

As for the other signs, we apply the negative relation between the level of the energy prices and the energy expenditure tax in the expressions obtained from the *pure analytical method*. In such a way, we can identify the positive and negative effects on the scrapping age and the investment in the long run. In what follows, we analyze the behaviour of the scrapping age along the BGP.

Optimal scrapping age

Taking the *pure analytical method* and rearranging the function F_1 (.), we get:

$$\frac{\partial \bar{T}}{\partial Z} = -\left(\frac{1}{\bar{T}} - \frac{\gamma \exp\{\gamma \bar{T}\}}{\exp\{\gamma \bar{T}\} - 1} + \frac{\gamma}{1-\alpha} \right)^{-1} \frac{1}{1-\alpha} \left(\frac{1}{\bar{P}_e} \frac{\partial \bar{P}_e}{\partial Z} + \frac{1}{1+Z} \right) \quad (17)$$

Here it is possible to distinguish two opposite effects of the energy expenditure tax on the scrapping age.
Direct effect: It is the effect of a modification in the energy tax on the scrapping age for a fixed level of energy prices:

$$\left. \frac{\partial \bar{T}}{\partial Z} \right|_{\bar{P}_e \text{ fixed}} = -\left(\frac{1}{\bar{T}} - \frac{\gamma \exp\{\gamma \bar{T}\}}{\exp\{\gamma \bar{T}\} - 1} + \frac{\gamma}{1-\alpha} \right)^{-1} \frac{1}{1-\alpha} \frac{1}{1+Z}$$

The term $(1/1 - \alpha)(1/1 + Z)$ is always positive because $0 < \alpha < 1$ and $0 < Z < 1$. Considering $\gamma > 0$, $\rho > 0$ and $\overline{T} > 0$, it is easy to prove that:

$$\left(\frac{1}{\overline{T}} - \frac{\gamma \exp\{\gamma \overline{T}\}}{\exp\{\gamma \overline{T}\} - 1} + \frac{\gamma}{1 - \alpha} \right)^{-1} > 0$$

Then, the direct effect has a negative effect on the scrapping rule; *i.e.*:

$$\partial \overline{T} / \partial Z \big|_{\overline{P}_e \, fixed} < 0$$

So, if the energy expenditure tax increases, the scrapping age is reduced for a fixed level of energy prices. The interpretation of this effect is clear. If the energy tax increases, the operation cost rises for a fixed level of energy prices. Consequently, firms decide to replace their equipment earlier. We can verify this explanation taking the scrapping and investment rule in the long run.

When the tax increases, firms can modify the decision about the scrapping age and investment. The net result is given by substituting the *scrapping rule* (14) into the *investment rule* (13). After some manipulations it yields:

$$(1 + Z)\overline{P}_e \left[\left(\frac{1}{\rho} - \frac{1}{\gamma + \rho} \right) \exp\{(\gamma + \rho)\overline{T}\} + \frac{1}{\rho} \exp\{\gamma \overline{T}\} + \frac{1}{1 + \rho} \right] = 1$$

When Z rises, firms compensate the higher tax by dropping the scrapping age for a fixed level of energy prices.

However, according to our simulation, this negative direct effect is overcome by an *indirect effect* of the energy expenditure tax on the scrapping age through the variation of the level of energy prices. This effect is described by the following expression:

$$-\left(\frac{1}{\overline{T}} - \frac{\gamma \exp\{\gamma \overline{T}\}}{\exp\{\gamma \overline{T}\} - 1} + \frac{\gamma}{1 - \alpha} \right)^{-1} \frac{1}{1 - \alpha} \frac{1}{\overline{P}_e} \frac{\partial \overline{P}_e}{\partial Z}$$

from equation (17). Since $\partial \overline{P}_e / \partial Z$ is negative, the indirect effect is positive. When the energy tax rises, the level of energy prices decreases. As a consequence, the operation cost of machines is reduced, and firms want to scrap their equipment later.

To summarize, we can conclude the following. When the tax on the energy expenditure of firms rises, the operation cost of machines increases. Then, firms decide to replace their equipment earlier (*direct effect*). However, this effect is overcome by the reduction in the level of energy prices, which is also produced by the increasing of the tax (*indirect effect*). Hence, the net effect of an increase of the energy tax on the scrapping age is positive $\partial \bar{T} / \partial Z > 0$. Therefore, our result gives theoretical evidence that an increase of an already high energy expenditure tax does not induce earlier replacement of machines; this is because that tax also modifies the level of energy prices.

Optimal investment

Investment is another important decision for firms, together with the scrapping age of machines. Here, we study how a tax on the energy expenditure of firms affects investment choice; *i.e.*, $\partial i^* / \partial Z$.

Differentiating the *scrapping rule* in the long run (14), and rearranging terms, we get:

$$\frac{\partial i^*}{\partial Z} = -i^* \left[\frac{1}{1-\alpha} \frac{1}{1+Z} + \left(\frac{1}{\bar{T}} + \frac{\gamma}{1-\alpha} \right) \frac{\partial \bar{T}}{\partial Z} + \frac{1}{1-\alpha} \frac{1}{\bar{P}_e} \frac{\partial \bar{P}_e}{\partial Z} \right] \quad (18)$$

Here we can distinguish a *direct effect* of the tax on investment and an *indirect effect* – through the scrapping age and the level of energy prices. For a fixed scrapping age and level of energy prices, we get the *direct effect*:

$$\left. \frac{\partial i^*}{\partial Z} \right|_{\bar{P}_e \text{ and } \bar{T} \text{ fixed}} = -i^* \frac{1}{1-\alpha} \frac{1}{1+Z} < 0$$

This effect is negative. If the tax increases, the operation cost of machines rises, too. Then, firms decide to invest less for a fixed scrapping age and level of energy prices.

However, there is an additional *indirect effect* from the variation of the scrapping age and the level of energy prices. This effect is given by the expression:

$$-i^* \left[\left(\frac{1}{\bar{T}} + \frac{\gamma}{1-\alpha} \right) \frac{\partial \bar{T}}{\partial Z} + \frac{1}{1-\alpha} \frac{1}{\bar{P}_e} \frac{\partial \bar{P}_e}{\partial Z} \right]$$

The sign of the *indirect effect* is undetermined because $\partial \bar{T} / \partial Z > 0$ and $\partial \bar{P}_e / \partial Z < 0$. Nevertheless, considering both effects together (*direct* and *indirect effect*), the net result is clear from the simulation: $\partial i^* / \partial Z < 0$. The explanation is the following. The *direct effect* reduces investment, because an increase of the tax raises the operation cost. However, this effect is not offset by the *indirect effect* of the scrapping age and the level of energy prices. Indeed, according to our simulation, the behaviour of the scrapping age reinforces the *direct effect*.

Final good output

The static comparative of the final good output is given directly from the static comparative of both the optimal scrapping age and the optimal investment considered before. The $\partial y^*/\partial Z$ is given by the equation of the final good output in equilibrium, $y^* = A(i^* \bar{T})^\alpha$. Differentiating that expression with respect to Z, we get:

$$\frac{\partial y^*}{\partial Z} = \alpha y^* \left(\frac{1}{i^*} \frac{\partial i^*}{\partial Z} + \frac{1}{\bar{T}} \frac{\partial \bar{T}}{\partial Z} \right)$$

From the previous sections, we know that $\partial \bar{T} / \partial Z > 0$ and $\partial i^* / \partial Z < 0$. According to our simulation, the effect of increasing an already high energy expenditure tax on the final good output is negative; the decrease of the optimal investment overcomes the later replacement of equipment.

CONCLUDING REMARKS

We focused our analysis on the long-run consequences of modifications in a tax on the energy expenditure of firms. The methodology developed here is a combination of numerical and analytical methods. We found that our model is very rich to capture the different elements that affect the long-run behaviour of our economy. In particular, we point out the usually forgotten issue of equipment replacement, which plays an important role in energy-saving technological change. One important consequence of considering such a replacement is that an increase of an already high energy expenditure tax does not induce earlier replacement of machines. The reason is that a tax on energy also modifies the level of energy prices.

Obviously, our analysis has some limitations. The main restriction here is the assumption of *exogenous* energy-saving technological progress. Indeed, it is clear that a tax on the energy expenditure of firms is going to affect technical progress. There are some studies in the literature concerning the importance of *endogenous* technological progress in these kinds of models. See, for example, Carraro, Gerlagh and van der Zwaan (2003) or Buonnano, Carraro and Galeotti (2003). In addition, since we considered a general equilibrium model, a good way to implement this idea is through an R&D sector (see Löschel, 2002 for a survey about technological change in economic models of environmental policy).

Finally, a second interesting extension could be the inclusion of an extraction sector. In this chapter, we assumed available energy supply to be exogenous. However, with an endogenous energy supply, the behaviour of the energy prices would be more realistic (especially in the short run).

APPENDIX

In this chapter, we write a program for Gauss to solve the static system of non linear equations given by the expressions (13), (14) and (16). The structure is simple. First, we assign values to the parameters of our model from the economic literature to get an optimal scrapping age and ratio optimal investment/gdp around, respectively, 16 years and 16 percent. After, we solve the static system of non-linear equations by a standard Newton-Raphson algorithm.

Table 11.1   Calibration

Parameter	A	α	ρ	γ	$\bar{e}_s$	Z
Value	15	1/3	0.05	6%	400	[0.8,0.9]

Table 11.2   Results

Derivative	$\partial \bar{P}_e / \partial Z$	$\partial \bar{T} / \partial Z$	$\partial i^* / \partial Z$	$\partial y^* / \partial Z$
Sign	−	+	−	−

NOTES

1. However, this regularity may not be true along the transition.
2. It could be considered as a lump-sum tax.
3. The available energy supply is a flow (exogenous) variable, for example. Here we do not explicitly treat an extraction sector.
4. Such an equilibrium is well known in the economic literature; for example, P.K. Bardhan (1966) and Boucekkine *et al.* (1997).

REFERENCES

Bailey, I. (2002), 'European environmental taxes and charges: economic theory and policy practice', *Applied Geography*, **22**, 235-251.

Bardhan, P.K. (1966), 'International trade theory in a vintage-capital model', *Econometrica*, **34**, 756-767.

Berndt, E. and D. Wood (1974), 'Technology, prices and derived demand for energy', *The Review of Economics and Statistics*, **57**, 259-268.

Boucekkine, R., M. Germain and O. Licandro (1997), 'Replacement echoes in the vintage capital growth model', *Journal of Economic Theory*, **74** (2), 333-348.

Boucekkine, R. and A. Pommeret (2004), 'Energy saving technical progress and optimal capital stock: the role of embodiment', *Economic Modelling*, **21**, 429-444.

Buonanno, P., C. Carraro and M. Galeotti (2003), 'Endogenous induced technical change and the costs of Kyoto Protocol', *Resource and Energy Economics*, **25**, 11-34.

Carraro, C., R. Gerlagh and B. van der Zwaan (2003), 'Endogenous technical change in environmental macroeconomics', *Resource and Energy Economics*, **25**, 1-10.

Dasgupta, P. and M.G. Heal (1974), 'The Optimal depletion of exhaustible resources', *Review of Economic Studies*, Symposium.

D'Autume, A. and P. Michel (1993), 'Endogenous growth in Arrow's learning by doing model', *European Economic Review*, **37**, 1175-1184.

Dixit, A. and J. Stiglitz (1977), 'Monopolistic Competition and optimum product diversity', *American Economic Review*, **67**, 297-308.

Hartwick, J.M. and N. Olewiler (1986), *The Economics of Natural Resource use*, New York: Harper Collins.

Hudson, E. and D. Jorgenson (1975), 'US energy policy and economic growth 1975-2000', *Bell Journal of Economics and Management Science*, **5**, 461-514.

Löschel, A. (2002), 'Technological change in economic models of environmental policy: a survey', *Ecological Economics*, **43**, 105-126.

Newell, R., A. Jaffe and R. Stavins (1999), 'The induced innovation hypothesis and energy saving technological change', *Quarterly Journal of Economics*, **114**, 941-975.

Pérez-Barahona, A. and B. Zou (2004), 'A comparative study of energy saving technical progress in a vintage capital model', *Resource and Energy Economics*. Forthcoming.

Smith, V.L. (1961), *Investment and Production: A Study in the Theory of the Capital-Using Enterprise*. Cambridge, Massachusetts: Harvard University Press.

Smulders, S. and M. de Nooij (2003), 'The impact of energy conservation on technology and economic growth', *Resource and Energy Economics*, **25**, 59-79.

Terborgh, G. (1949), *Dynamic Equipment Policy*, Machinery and Allied Products Institute, Washington D.C: McGraw-Hill.

12. Oil shocks and the business cycle in Europe

Carlos de Miguel, Baltasar Manzano and José M. Martín-Moreno

INTRODUCTION

The effects of oil price changes on economic activity have been widely studied. An increase of oil prices tends to reduce the level of economic activity, given its implications on the evolution of important macroeconomic variables and due to the strong dependence of western economies on this input.

Recently, we have witnessed a substantial increase in the prices of oil, derived mainly from the high demand of emerging economies (China among others) and from the political and economic situation of Iraq. In this framework, the economic problems derived from trends in oil prices appear again, not only with regard to the situation of the markets, but also to their possible macroeconomic effects. Tensions in the oil markets have returned, and we are perhaps entering a period that some experts call the end of the cheap oil age. The ghost of old crises seems to be reappearing, stirring a fear of recessions in the main economies that integrate the European Union.

As it is very well-known, for economies so dependent on oil, increases in the price of the barrel have different adverse effects. In the short run, they involve a reduction of GDP and an increase of the inflation rate. In the medium and long run, industrial production is affected, consumption decreases due to the fall of purchasing power and investment also falls, affecting the cyclical position of the economy and citizens' welfare. Consequently, with oil prices between 40 and 50 dollars per barrel, the economic implications should be studied.

However, to be able to carry out a rigorous analysis of what this new oil crisis in the European setting can represent, and to make an accurate analysis of economic effects, we must situate ourselves in a certain time frame. Thus, in this chapter, we analyze the effects of the increase in oil prices since the first crisis (70s), on economies so dependent on oil as the Europeans.

The goal of the chapter is twofold. Firstly, we analyze the importance of the source of shocks for both the size and the shape of aggregate fluctuations in those economies during the last three decades. Secondly, we quantify the effects of the changes in relative oil prices on welfare in these economies.

The effects of oil price shocks on industrialized economies have been widely acknowledged in economic literature. Pindyck (1979), Hamilton (1983) and Olson (1988) suggest that these shocks affect growth as well as the business cycle, thereby becoming an additional source of economic fluctuation. There is extensive empirical literature that offers evidence of an asymmetric relationship between oil prices and aggregate economic activity (see Mork, 1989 and 1994). Kim and Loungani (1992) and Finn (1995) analyze the role of energy price shocks using real business cycle models in closed economies, focusing on the US. These authors find that such shocks offer very little help in explaining the aggregate fluctuations in the economy in question. However, De Miguel, Manzano and Martín-Moreno (2003) show that, in a small open economy framework, oil price shocks are very important when explaining aggregate fluctuations.

This chapter presents a real business cycle model for the different European economies. The model used is a standard general equilibrium model of a small open economy in which oil is included as an imported productive input. The relative oil price as well as the real interest rate is assumed to be set in international markets, so we consider a small open economy in the sense of taking those prices as given. Oil price shocks are the only source of fluctuation considered. Therefore, although the economy is hit by many shocks, our analysis is conditional on a single shock. Thus, this analysis would allow us to verify the extent to which oil price shocks can account for aggregate fluctuations in European economies.

The results show that oil shocks can account for a significant percentage of GDP fluctuations in many of those countries, although the explanatory power is smaller for others. This wide range of variation can be explained by differences in the strength of monetary policies that affect relative oil prices in each country. In addition, the model reproduces the cyclical path of the European economies in periods of oil crisis. Finally, we show that increases in the relative price of oil had a negative effect on welfare, particularly in southern European countries, which are historically associated with a lax monetary policy during oil crises.

THE MODEL

The model described in this section is a stochastic dynamic general equilibrium model of a small open economy populated by a large number of infinite-lived households, and firms that need to import oil to produce a consumption good. The basic structure of the model, in terms of preferences and technology, is similar to the De Miguel, Manzano and Martín-Moreno (2003) structure. The general features of the model are the following.

The production of the final good, y_t, requires the use of labor, n_t, capital, k_t and energy, e_t. The production technology of firms is described by a nested CES function with constant returns to scale:

$$F(n_t,k_t,e_t) = n_t^\theta \left[(1-a)k_t^{-\upsilon} + ae_t^{-\upsilon} \right]^{-\frac{1-\theta}{\upsilon}}, \tag{1}$$

where θ is the labor share and the parameter υ is equal to $(1-s)/s$, where s is the elasticity of substitution between capital and energy.

The economy's resource constraint for period t is given by:

$$c_t + i_t + xn_t = y_t, \tag{2}$$

where c_t is private consumption, i_t is investment and xn_t are net exports. Capital, k_t, accumulates according to the law of motion:

$$i_t = k_{t+1} - (1-\delta)k_t + \Phi(k_t,k_{t+1}), \tag{3}$$

where δ is the depreciation rate and Φ is a capital adjustment cost function which we assume to be quadratic:

$$\Phi(k_t,k_{t+1}) = \frac{\phi}{2}\left(\frac{k_{t+1} - k_t}{k_t} \right)^2, \tag{4}$$

The representative firm solves:

$$Max \ F(n_t,k_t,e_t) - w_t n_t - r_t k_t - p_t e_t, \ \forall t \tag{5}$$

where w_t is the wage, r_t the capital rate of return and p_t the relative oil price. In equilibrium, marginal productivities are equal to input prices:

$$w_t = F_{tn}, \quad p_t = F_{e_t} \quad \text{and} \quad r_t = F_{k_t}.$$

The relative oil price follows a stationary stochastic process:

$$\ln p_t = \overline{p} + \rho \ln p_{t-1} + \varepsilon_t, \quad \varepsilon_t \sim N(0, \sigma_p), \quad |\rho| < 1. \qquad (6)$$

We assume that the individuals can buy or sell an international asset, b, at an exogenous international interest rate, r^*. The evolution of net exports is dictated by:

$$xn_t = p_t e_t + b_{t+1} - (1 + r^*) b_t, \qquad (7)$$

where $p_t e_t$ are oil purchases.

Consumers maximize the expected value of lifetime utility subject to their budget constraint:

$$Max \ E_0 \left\{ \sum_{t=0}^{\infty} \beta^t \frac{1}{1 - \sigma} \left[\left(c_t - \psi n_t^v \right)^{1-\sigma} - 1 \right] \right\}$$

$$s.t. \quad c_t + k_{t+1} - (1 - \delta) k_t + \Phi(k_t, k_{t+1}) + b_{t+1} = w_t n_t + r_t k_t + (1 + r^*) b_t, \qquad (8)$$

where $0 < \beta < 1$ is the subjective rate of intertemporal discount, $\sigma > 0$ is the parameter of relative risk aversion, v is one plus the inverse of the intertemporal elasticity of substitution of labor supply and ψ is a positive parameter.

The conditions that solve the consumer's problem are the following:

$$U_{n_t} + U_{c_t} w_t = 0, \qquad (9)$$

$$U_{c_t} \left(1 + \phi \frac{k_{t+1} - k_t}{k_t^2} \right) = \beta E_t \left\{ U_{c_{t+1}} \left(1 - \delta + r_{t+1} + \phi \frac{k_{t+2} - k_{t+1}}{k_{t+1}} \frac{k_{t+2}}{k_{t+1}^2} \right) \right\}, \qquad (10)$$

$$U_{c_t} = \beta E_t \left\{ U_{c_{t+1}} (1 + r^*) \right\}, \qquad (11)$$

$$c_t + k_{t+1} - (1 - \delta) k_t + \Phi(k_t, k_{t+1}) + b_{t+1} = w_t n_t + r_t k_t + (1 + r^*) b_t. \qquad (12)$$

PARAMETER VALUES

We now briefly describe our procedures for selecting parameter values listed in Table 12.1. We follow the standard real business cycle literature in using steady-state conditions to find parameter values matching average values observed in the data, while other parameters will be equal to standard values used in the literature. The model is calibrated to reproduce average values of the European Union in annual data from 1960–2003, before the unification of 2004 (EU-15). The main sources for the data used are the AMECO Database from Eurostat and International Energy Agency (IEA) Statistics.

The depreciation rate of capital is obtained from equation (3) in steady state, $\delta = i/k$, where i/k is the average value of EU-15 data, while the discount factor β was set by using equation (11) in steady state, $1 = \beta(1+r^*)$, so as to give a real annual interest rate of 4 percent.

Table 12.1 Parameters of the economy

Preferences		
Subjective Discount Rate	β	0.96
Parameter of the Utility Function	ψ	1.53
Risk Aversion	σ	1.001
Parameter of the Utility Function	v	1.7
Technology		
Labor Share	θ	0.64
Rate of Depreciation	δ	0.06
Parameter of the Production Function	υ	0.7
International Interest Rate	r^*	0.04

The value of a, representing the importance of oil respect capital in the production function, is obtained from the first order conditions of the firm's problem in the steady state, $a = 1/[(e/k)^{-(\upsilon+1)}((r^* + \delta)/p)+1]$ where e/k and p represent average values over the sample, and υ is borrowed from Kim and Loungani (1992).

The value of the parameter ψ is chosen from (9) in the steady state: $\psi = \theta(y/n)/vn^{(\upsilon-1)}$, assuming that the productive time is 5476 hours per year. The output per worker, y/n and the labor share, θ, represent averages for the EU-15 economies.

The parameter of the adjustment costs ϕ is calibrated so the variability of investment relative to output in data would be reproduced by the model.

The remaining parameters are chosen in conformity with earlier studies. The parameter of the utility function v is taken from Greenwood *et al.* (1988), while the risk aversion parameter σ is obtained from Mendoza (1991).

Finally, the stochastic process parameters of relative oil prices for each country of the EU-15 are estimated, in domestic currency, from equation (6). Table 12.2 reports the results of the estimation.

Table 12.2 Oil price process

	Constant $\overline{p}$	Persistence Coefficient ρ	Standard Deviation σ_p
Portugal	0.63	0.82	0.32
Spain	0.63	0.81	0.30
Greece	0.77	0.81	0.29
Italy	1.16	0.80	0.32
France	0.01	0.75	0.32
United Kingdom	-0.30	0.83	0.32
Ireland	-0.48	0.77	0.32
Germany	-0.25	0.78	0.32
Belgium	0.44	0.76	0.31
Luxembourg	0.43	0.77	0.30
Netherlands	-0.29	0.74	0.31
Austria	0.18	0.79	0.32
Sweden	-0.30	0.82	0.33
Finland	-0.02	0.77	0.30
Denmark	-0.40	0.78	0.31

OIL SHOCKS AND EU-15 AGGREGATE FLUCTUATIONS

In this section, we test how accurately the model driven by oil price shocks can fit the business cycle of the EU-15 between 1970 and 2003. We start by running simulations, including for each country the corresponding stochastic process for the relative oil price presented in Table 12.2. Such an experiment allows us to compare the actual GDP data with the corresponding fluctuations of the output in the model, obtaining the percentage of GDP volatility that can be explained by the model with oil shocks. Thus, we can explore to what extent output fluctuations at each European country could be generated by oil shocks. The simulation results are summarized in Table 12.3.

Table 12.3   Comparison between the predictions of the model and the data

	Data GDP (standard deviation)	Model Explanation of Output volatility
Portugal	5.51%	30%
Spain	3.47%	37%
Greece	4.07%	42.6%
Italy	3.17%	41.8%
France	2.06%	22%
United Kingdom	1.51%	10.6%
Ireland	2.36%	8.9%
Germany	2.69%	10%
Belgium	2.02%	19.8%
Luxembourg	2.82%	15.6%
Netherlands	2.16%	16.7%
Austria	2.14%	28.5%
Sweden	1.30%	33.1%
Finland	4.08%	13.2%
Denmark	1.38%	30.4%

The relative price of oil in each country is obtained as the ratio between the price of the Brent barrel expressed in domestic currency and the corresponding GDP deflator. Therefore, the differences of relative oil prices between countries arise either from inflation or from the exchange rate. In this sense, both the exchange rate and the monetary policies would have been the main tool to accommodate the effects of oil crises in each country.

In light of the simulation results, we can consider several groups of countries. The first group includes Portugal, Spain, Greece and Italy, that is, countries with large GDP fluctuations, where the role of oil shocks is remarkable, responsible from 30 percent to 42 percent of total output fluctuations in Portugal and Italy, respectively. Those results are consistent with the common view about southern European countries' excessively lax monetary policy, which led to more difficulties in accommodating oil shocks.

There is another group of countries (Austria, Denmark and Sweden) in which oil disturbances also play an important role, explaining around 30 percent of output fluctuations, although their output is quite less volatile.

The main group includes the rest of the EU-15 countries, in which the contribution of oil shocks to explain GDP fluctuations scores from 10 percent to 20 percent. That group includes countries like France, UK, Netherlands, Belgium, Luxembourg and Germany, whose central banks have implemented stronger monetary policies to face the oil crises of the 70s.

Therefore, as it was pointed out above, the role of monetary policies has been crucial in accommodating the oil shocks that hit western economies in the last three decades.

Oil shocks are an important source for explaining aggregate fluctuations across Europe, although with a wide range of variation. But it is also important to test whether the introduction of oil shocks can mimic the business cycle shape. In order to do that, we simulate the model with the actual path of the relative price of oil as the only source of fluctuation. The analysis allows us to verify the extent to which oil prices actually give rise to the business cycle path of the EU-15 countries.

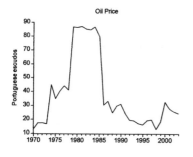
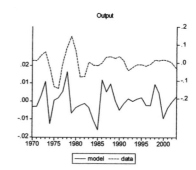

Figure 12.1   Portugal

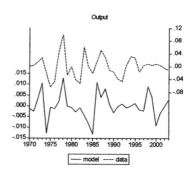

Figure 12.2   Spain

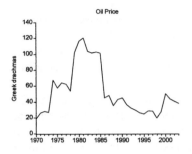

Figure 12.3   Greece

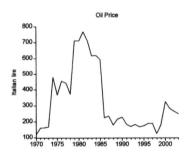
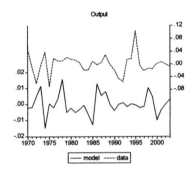

Figure 12.4   Italy

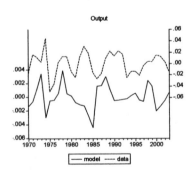

Figure 12.5   France

Figure 12.6   United Kingdom

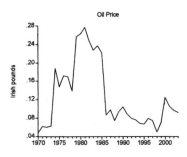
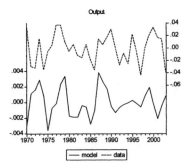

Figure 12.7   Ireland

Figure 12.8   Germany

Figure 12.9   Belgium

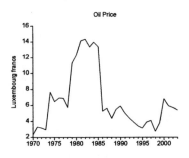
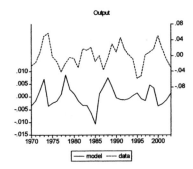

Figure 12.10   Luxembourg

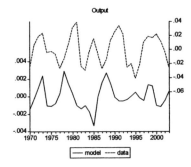

Figure 12.11   Netherlands

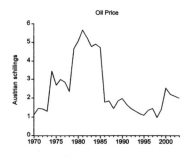
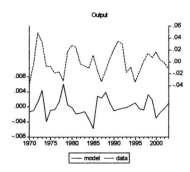

Figure 12.12   Austria

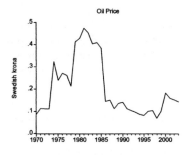
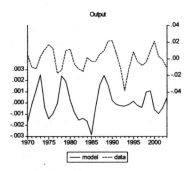

Figure 12.13   Sweden

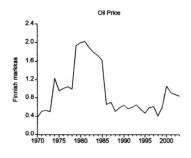
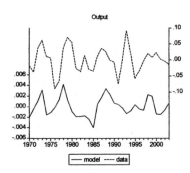

Figure 12.14   Finland

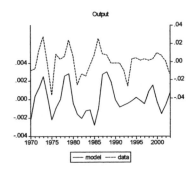

Figure 12.15   Denmark

Figures 12.1–15 represent the evolution of the relative price of imported oil, and the comparison between the evolution of the actual GDP and the output path simulated by the model for each country. Given that we focus on the business cycle, both series have been detrended using the Hodrick-Prescott filter.

The chart of the relative oil prices roughly shows the same path for all countries analyzed. The first great oil crisis was triggered by the Arab–Israeli Yom Kippur War when the OPEC imposed an oil embargo on western countries, thereby dramatically increasing oil prices at the beginning of 1974. The increase had immediate effects on the evolution of the output in the model as well as on the GDP data, though with some delay in several countries. Therefore, the model satisfactorily reproduces the negative effect on output during the first oil crisis. The next period which calls for our attention is the second oil crisis, which takes us back to 1979 and was triggered by the Iranian Revolution, causing the relative price of oil to double in less than a year. Again, the increase in oil prices reduces the output in both the model and the economy in most countries, which is accurately depicted by the model. Some countries responded to the oil crisis with some delay (Germany, Belgium, Netherlands, Austria and Italy). At the beginning of 1986, the oil market collapsed, bringing about a significant drop in oil prices. This situation affected output in a positive way, which quickly recovered. The model reacts instantly when confronted with the fall in oil prices, while the recovery in the data begins some periods later in most countries.

Summing up, the model guided by changes in oil prices is able to mimic the shape of the output fluctuations in episodes of dramatic changes in the oil market, especially for the two crises in the 70s and the situation in the mid-

80s. However, when oil market conditions were stable, as they became in 1986, there are other disturbances that explain aggregate fluctuations, so the model driven by oil shocks fails to reproduce the cyclical path of the countries analyzed.

The results show the vulnerability of small open economies heavily dependent on imported oil when they are confronted with large changes in the conditions that control the international oil market. This highlights the importance of considering the behavior of relative oil prices when analyzing the business cycle.

OIL PRICE SHOCKS AND WELFARE

The previous section stresses the influence of oil shocks on both the size and the shape of aggregate fluctuations. Nevertheless, oil shocks affect not only GDP fluctuations but also welfare. As De Miguel, Manzano and Martín-Moreno (2003) emphasized, those effects are particularly important in the framework of a small open economy dependent on oil imports because of no domestic oil production, and the possibilities of substituting this input with capital are limited due to the complementarity between capital and oil. In this section, we evaluate the loss of welfare occasioned by oil price increases registered from the first half of the 70s until the second half of the 80s. The relative price in the last quarter of 1973 was very similar to the value registered in the third quarter of 1986 in each country; therefore, we can consider the intermediate period as a temporary price increase whose welfare cost we calculate.

The welfare loss of oil crises is estimated by the welfare cost derived from the increase in oil prices with respect to the initial situation. This cost is defined at each moment as the percentage of the increase in consumption that an individual would require to enjoy the same level of welfare with respect to the starting point. The welfare cost at each moment is calculated through the x variable that solves the following equation:

$$\overline{U} = U\left[\tilde{c}(1+x), \tilde{n}\right], \tag{13}$$

where $\overline{U}$ is the level of utility reached at the initial situation, in this case 1973, while $\tilde{c}$ and $\tilde{n}$ represent consumption and hours worked at each moment, which provide a level of welfare U. Thus, the product $\tilde{c}x$ indicates the total increase in consumption required to restore the initial level of welfare. This welfare cost measure is usually expressed as a percentage of output.

Table 12.4 Welfare cost: 1974–1985

	Average cost per period (percent of the Output)
Portugal	8.14%
Spain	5.98%
Greece	7.93%
Italy	6.97%
France	1.74%
United Kingdom	0.69%
Ireland	0.74%
Germany	1.12%
Belgium	3.59%
Luxembourg	3.67%
Netherlands	1.04%
Austria	2.33%
Sweden	0.87%
Finland	1.51%
Denmark	0.79%

The results, summarized in Table 12.4, show that the welfare cost from oil crises was very different between countries. Again, southern European countries, with a lax monetary policy, had a larger welfare cost, ranging from 6 percent of output in Spain to more than 8 percent in Portugal, so consumers should have been compensated in each period with a significant fraction of GDP in terms of consumption in order to make up for the loss in welfare derived from the different oil crises. The welfare cost in the rest of the countries was moderate (Belgium and Luxembourg) or even small. This gives us an idea of the significant amount of the loss in welfare brought about by the different oil crises for a small open economy, and the importance of managing monetary policy to accommodate such kinds of shocks.

CONCLUSIONS

In this chapter, we have analyzed the effects of oil price shocks on the business cycle of the EU-15 countries. The model used for this analysis is based on the standard dynamic general equilibrium model for a small open economy in which oil is included as an imported productive input. The price of oil and the interest rate are assumed to be set by international markets. The calibration of the parameters of the model is carried out by taking data from the EU-15 economies during the period 1960–2003.

The results show that oil shocks can account for a significant percentage of GDP fluctuations in many of those countries, but the explanatory power is quite smaller for others. That wide range of variation can be explained by differences in the strength of monetary policies. In addition, the model reproduces the cyclical path of the European economies in periods of oil crisis. Finally, we have shown that the increases in the relative price of oil had a negative effect on welfare, particularly in southern European countries, which are historically associated with a lax monetary policy during oil crisis.

NOTE

* Financial support from Spanish Ministry of Education and Science and FEDER through grant SEJ2005-03753/ECO and from Xunta de Galicia (PGIDIT03PXIC30001PN, PGIDIT03CSO30001PR) is gratefully acknowledged.

REFERENCES

De Miguel, C., B. Manzano and J.M. Martín-Moreno (2003), 'Oil price shocks and aggregate fluctuations', *The Energy Journal*, **24**(2), 47-61

Finn, M. (1995), 'Variance properties of Solow's productivity residual and their cyclical implications', *Journal of Economic Dynamics and Control*, **19**, 1249-1281.

Greenwood, J., Z. Hercowitz and G. Huffman (1988), 'Investment, capacity utilization and the real business cycle', *American Economic Review*, **78**, 971-987.

Hamilton, J. (1983), 'Oil and the macroeconomy since World War II', *Journal of Political Economy*, **91**, 228-248.

Kim, I. and P. Loungani (1992), 'The role of energy in real business cycle models', *Journal of Monetary Economics*, **29**, 173-189.

Mendoza, E.G. (1991), 'Real business cycles in a small open economy', *American Economic Review*, **81**, 791-818.

Mork, K.A. (1994), 'Business cycles and the oil market', *The Energy Journal*, Special issue, The Changing World Oil Market, 15-38.

Mork, K.A. (1989), 'Oil and the macroeconomy when prices Go Up and Down: An extension of Hamilton's results', *Journal of Political Economy*, **97**, 740-744.

Olson, M. (1988), 'The productivity slowdown, the oil shocks, and the real cycle', *The Journal of Economic Perspectives*, **2**, 43-70.

Pindyck, R.S. (1979), *The Structure of World Energy Demand*, Cambridge MA: The MIT Press.

13. Energy transitions and policy design in a GPT setting with cyclical growth through basic and applied R&D

Adriaan van Zon and Tobias Kronenberg

INTRODUCTION

The current bad news about the adverse impact of human activity on the environment[1] stresses the need to reconcile further economic growth with environmental protection. Since most environmental damage is caused by the consumption and distribution of energy, a decoupling of growth from pollution will require a massive reorganisation of the energy system, possibly leading towards a hydrogen economy (Rifkin, 2002). We are interested in finding out how economic policy may help to promote the transition from the current energy system to a sustainable alternative. It seems to us that the transition from the one fuel paradigm to the other is closely related to what Bresnahan and Trajtenberg (1995) and Helpman (1998) refer to as transitions between 'General Purpose Technologies' (henceforth GPT's), which come into existence as the result of some drastic innovation. Drastic innovations are usually contrasted with *incremental innovations* that extend/build upon the basic idea embodied in the drastic innovation. Both types of innovations taken together define, in our terminology, a so-called 'technology family', with the drastic innovation serving as its core.

In order to incorporate drastic and incremental innovations into one model,[2] we use a multi-level Ethier[3] production function employing effective capital that consists of several GPT's that are used simultaneously and that are each made up of a core innovation and its accompanying peripheral innovations. At both levels there will be love of variety effects (cf. van Zon and Yetkiner, 2003), and researchers will have to engage in either basic research that is aimed at finding new cores of new GPT's, or applied R&D geared at expanding the field of application of the core of a GPT through the addition of 'peripheral' innovations. Thus, each successful basic R&D project gives rise to follow-up applied R&D projects.

The model incorporates the uncertainty associated with drastic innovations by drawing the parameters that determine a technology's characteristics from a random distribution. Thus, a new technology may result in a successful GPT with many practical applications, a complete failure, or anything in between these two extremes. Finding a new core of a GPT enables other researchers to begin applied R&D in the new technology.

For our present purposes, we extend the ZFK-model in several ways. *First*, we allow for different types of fuels, a carbon-based one and a non-carbon-based one, and assume that carbon-based fuels generate adverse environmental externalities. *Secondly*, we introduce two types of technical progress giving rise to the development of the cores and peripherals of new carbon-based GPT's or non-carbon-based GPT's. *Third*, we add emissions thus defining (albeit in a very simplistic way) part of the trade-off between growth and environmental quality. Using the extended ZFK-model, we focus on the interplay between different types of R&D in a stochastic setting. The latter forces us to use a simulation version of the model. In addition to this, we have simplified those parts of the model that are not directly R&D related.[4] The R&D features of the ZFK-model, on the other hand, are such that we get cyclical growth through the interplay of different types of R&D. This is due to our assumption that the expansion of GPT's through applied R&D eventually runs into diminishing returns, thus inducing a shift towards basic R&D, which in turn generates new possibilities for applied R&D. In short, the ZFK-model emphasizes the Schumpeterian aspects of growth, but also the role of R&D in generating cyclical growth, as well as the uncertainties surrounding R&D activity, apart from disaggregating R&D activity itself into basic R&D and as many applied R&D processes as there are GPT's.

The extended ZFK-model contributes to the literature in many different ways. *First*, it introduces asymmetries in the intermediate goods market that arise from our assumption that the contribution of the latest peripherals/intermediates to a GPT decreases as more and more peripherals are added to the GPT.[5] *Secondly*, we have multiple GPT's that are active at the same time, as in real life, and as opposed to more standard GPT models. *Third*, we have cyclical growth that does not depend on the reallocation of homogeneous labour between production and R&D activities, as is the case in the GPT approach by Bresnahan and Trajtenberg (1995), or in more recent work focusing on the clustering of R&D activity against a GPT-background that explains booms and busts in the business cycle (Francois and Lloyd-Ellis, 2003, or in Matsuyama, 1999). Rather than this being a weakness of our model, we feel that it is actually a strong feature, as it is hard to imagine that the 'eye-hand-coordination' in the words of Romer (1990) is suited in practice to perform R&D tasks, and the other way around. *Fourth*, our model elaborates on the role of basic and applied R&D mechanisms in the growth process. It shows that the influence of these two R&D types on the long-run growth process is very different indeed. *Fifth*, the model shows that in a competitive equilibrium, the allocation of researchers between basic and applied R&D will generally be inefficient. We will also show that this inefficient allocation of R&D may give rise to situations in which the introduction of a carbon tax may yield a double dividend. But in addition to this, our simulations also show that the effect of any policy scheme crucially depends on the existing technology structure causing policy making to become highly path-dependent. If, for example, we are in a situation where carbon-based technologies form the dominant paradigm, then a carbon tax will – as a side-effect – move R&D resources away from applied R&D on existing technologies towards basic R&D on new technologies, and this side-effect will reduce the market failure in the R&D sector. Thus the carbon tax alone tends to reduce carbon-based fuel consumption and increase growth. Tax recycling as a subsidy on non-carbon-based fuels will reinforce the carbon reduction, and tax recycling as a subsidy on basic R&D will reinforce the growth effect, and depending on preferences, any of these schemes could be desirable.

The organization of the rest of this chapter is as follows. In section 2 we provide a short summary of the most important features of the ZFK-model, and our energy extensions of that model. Section 3 shows the base run simulation results, and section 4 shows the results of some illustrative simulation exercises that were done using the model. Section 5 concludes.

THE ZFK-MODEL IN OUTLINE

General Overview

The organization of the extended ZFK-model resembles that of the Romer (1990) model. The model consists of a perfectly competitive final output sector that combines labour and effective capital to produce output. Effective capital is an aggregate of individual GPT's that in turn consist of intermediates supplied to the final output sector under imperfectly competitive conditions. Intermediates are built according to the specifications in the blueprints obtained from the R&D-sector that uses the market value of these blueprints to pay for the R&D labour and R&D entrepreneurship resources used in producing them.

The most important problem addressed in the ZFK-model is the 'spontaneous' arrival of new GPT's as the result of basic R&D, and the subsequent expansion of that GPT through applied R&D. Both types of R&D are profit-incentive driven, as is usual in new growth models *à la* Romer (1990) and Aghion and Howitt (1992, 1998). In our GPT-setup, the most important components of a GPT are invented first, while the invention of additional components through applied R&D become less and less profitable. The latter gives rise to decreasing returns to variety within a GPT, thereby increasing the relative attractiveness of finding the core of a new GPT rather than continuing the expansion of existing GPT's by finding still newer peripheral components.[6] However, it may well be the case that the core of a new GPT is not promising at all. Then our setup implies that the R&D sector will devote most of its resources trying to find a still newer core rather than producing peripheral inventions for a failing GPT.[7] Thus we get cyclical growth patterns, implied by inherent technological expansion limits, rather than by the reallocation of resources between the final output sector and the R&D sector, as one usually finds in GPT models.[8]

The R&D sector

We assume that each innovation, whether basic or applied, is the result of innovative activities in the R&D sector. Which type of R&D is done depends on the profitability of adding peripherals to an already existing technology (for which the core already exists) versus creating a completely new technology (for which a new core is required). Our motivation for considering different R&D processes for peripherals and cores is our perception that the invention of a technology core requires 'something more fundamental' than the further development of a technology by adding peripherals. We capture this difference by differentiating their contribution to total production. However, we also feel that finding a (core of a) potential GPT is subject to more uncertainty than finding a peripheral once a new technological 'proto–paradigm' has arrived in the form of a core of a potential GPT. We model this by assuming that the corresponding R&D process itself is uncertain, first because it is not possible to predict with complete certainty the exact arrival time of the core of a new GPT, and secondly because it is not possible to predict with complete certainty the actual characteristics of a potential GPT.[9] In order to model the uncertainty associated with basic research, we assign random values drawn from a uniform distribution to the characteristics of each GPT; these characteristics are unknown until the core of that GPT is actually invented. Applied R&D can only begin after the core has been introduced. At this point in time, the GPT characteristics become publicly known, and for reasons of simplicity, we assume that there is no uncertainty regarding the characteristics of new peripherals added to the GPT through applied R&D.[10] Of course, there is still a random element in applied R&D because the actual arrival of an applied R&D invention depends on random draws from a Poisson distribution.

Even though we explicitly distinguish between basic and applied R&D, we do model both types along very similar lines. In both cases, R&D gives rise to innovations that arrive according to a Poisson probability distribution. As in Aghion and Howitt (1992), we assume that the level of R&D activity directly and positively influences the arrival rate of innovations. But unlike Aghion and Howitt, we assume that the marginal productivity of R&D is decreasing, which in turn allows us to find the allocation of R&D workers over all different (basic and applied) R&D processes.

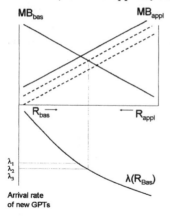

Figure 13.1   Allocation of R&D activities

Figure 13.1 shows how this allocation process works. The horizontal axis represents the (given) total number of R&D workers, who are either working in basic R&D or applied R&D. Moving to the right, the ratio of basic R&D to applied R&D increases, and the other way around. The two curves depict the value marginal product (VMP, for short) of employing basic and applied R&D.[11] Since we have assumed decreasing marginal R&D productivity, the curves are downward-sloping. The point of intersection of both VMP-curves determines the profit maximizing allocation of researchers, as both profits (and R&D wages if R&D workers are paid their value marginal product, as we assume) cannot be increased by moving from the low VMP R&D activity into the high VMP R&D activity.

If, in this setting, applied R&D is successful, a new peripheral is developed, and we know that the next peripheral will be less productive. Thus, the VMP-curve associated with applied R&D shifts downward. At the new point of intersection, more researchers are working in basic R&D than before. The bottom half represents the R&D 'production function'. It shows the arrival rate of new GPT's as a function of the number of active basic R&D workers. We can see that as more and more researchers move into basic R&D, the arrival rate increases, and the arrival of the next GPT becomes increasingly more likely. When the next GPT with a lot of development potential actually arrives, new possibilities for applied R&D arise. The VMP curve of applied R&D shifts up again, and the allocation of R&D workers changes in favour of applied R&D. Thus, the alternating arrivals of GPT cores and peripherals generate cycles in the R&D sector, with researchers switching between the two R&D activities, thus generating cyclical growth.

Closing the Model

We have introduced energy into the ZFK-model by means of the following modifications. *First* we distinguish between two types of GPT's that use either carbon-based or non-carbon-based fuels. *Secondly*, the marginal costs of the components of these GPT's now also include carbon- and non-carbon-based fuel costs, next to capital cost. In fact, we assume that fuel-capital ratios at the component level are fixed.[12] However, these fuel-capital ratios may differ between GPT's as this is another random GPT-characteristic.

For environmental quality, we simply assume that the degree of environmental degradation is proportional to the flow of carbon emissions. To simplify matters even more, we assume that pollution does not accumulate. This means that if carbon-based fuel consumption would be abolished altogether, the environment would return to its maximum quality immediately, which is obviously not a realistic assumption. For ease of exposition, we assume that the consumption of non-carbon-based fuels does not harm the environment at all.

Finally, we assume that fuel prices and the interest rate are exogenously determined.

Base run simulation results

In order to illustrate how the model works, we first present the results of a base run, using a set of arbitrary parameter values. Figure 13.2 shows how the number of available GPT's rises over time. We assume that the economy starts with just one carbon-based and one non-carbon-based GPT. Figure 13.3 shows the number of peripherals that are developed for each GPT. Interestingly, some GPT's develop only very few peripherals or none at all. These are 'failed' GPT's consisting of only a core with few or no peripherals. But there are also 'true' GPT's such as C03 or N04, which develop dozens of peripherals.[13] We see these 'failed' GPT's and 'true' GPT's in our *ex-post* perspective, but the researchers who developed them were seeing each of them as potential GPT's, because they could only guess at their true characteristics *ex ante*.

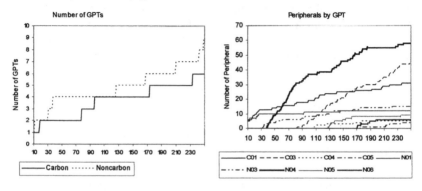

Figure 13.2 Availability of GPTs   *Figure 13.3 Peripherals by GPT*

One of the defining characteristics of a GPT is that it affects a large share of the economy. In order to measure this feature, we show the relative contribution to output (through the effective capital stock Ke) of each individual GPT.[14] The curves in Figure 13.4 show a cyclical pattern, which is consistent with some long-wave views on economic development, as in Freeman and Pérez (1988), for example, who identify five Kondratieff waves in economic history since the late eighteenth century that are characterised by the dominant GPT of their times. Note that in contrast to many long wave theories, there is nothing mechanistic in the coming and going of Kondratieffs in our model. In our model, every GPT will at some time run out of further extension possibilities, and the search for a new GPT begins, but the length of these 'long waves' is endogenously determined and highly variable.

Figure 13.5 shows how the dominance of a certain technology determines the economy's fuel mix. During the 'N03 Kondratieff', non-carbon-based fuel consumption is much higher than carbon-based fuel consumption. During the transition to C03, however, non-carbon-based fuel consumption levels off, while carbon-based fuel consumption quickly rises. At the end of the simulation period, the economy consumes a balanced mix of both fuels. There is no long-run tendency towards either fuel, because we assume that carbon-based fuel technologies are intrinsically just as productive as non-carbon-based fuel technologies. Thus, the R&D sector has no reason to concentrate on either fuel, and over the long run the economy can be expected to develop just as many carbon-based GPT's as non-carbon-based GPT's.

Figure 13.6 shows the allocation of R&D workers between basic and applied R&D. We have chosen the number of researchers – plotted on the vertical axis – to be equal to five. Just as in Figure 13.3, we clearly observe cycles in R&D. Whenever an attractive GPT is invented, researchers move into applied R&D on the new GPT. In year 31, for instance, the freshly invented N03 absorbs almost all of the R&D workers, who are busily developing peripheral applications based on that GPT. As the extension possibilities of N03 are being exploited, further R&D on N03 becomes less attractive, and researchers return to doing basic R&D.

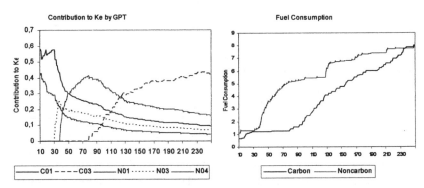

Figure 13.4   Contribution to effective   Figure 13.5   Technologies and fuel
capital stock by GPT   mix

The arrival of successful GPT's generates cycles in the growth rate of the economy as we can see in Figure 13.7, which shows the growth rate of real disposable income. Successful GPT's have two effects on growth. First, the invention of the GPT core raises output directly because of the Love of Variety production structure. Secondly, and more importantly, a new GPT can raise the average growth rate over a period of several decades, as researchers are exploiting the new possibilities for applied R&D. The actual growth impact of new GPT's, of course, depends on the intrinsic characteristics of the GPT, which in turn determine its pervasiveness, given the 'general GPT environment'.[15] In the extreme case, i.e., a total failure with no peripherals at all, we have only a 'growth-spike' due to the Love of Variety effect.

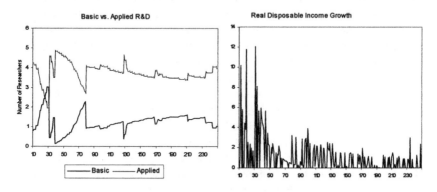

Figure 13.6   Allocation of R&D workers   Figure 13.7 Growth rate of real between basic and applied R&D

As regards the development of welfare, we have experimented with a CES function incorporating environmental quality and real disposable income as arguments. Generally speaking, we have found that as consumption and environmental quality become better substitutes, output growth also results in welfare growth, as increased consumption possibilities can more than make up for decreased environmental quality. In the extreme case where consumption and environmental quality would be perfect substitutes, it is always possible to make up the utility loss from environmental degradation through higher consumption. For low values of the elasticity of substitution between consumption and environmental quality, we find that the effect of environmental degradation may be so large that utility actually starts falling at some point in time. The reason for this is that growth reduces the marginal utility of consumption *vis-à-vis* that of environmental quality, making it increasingly difficult to raise welfare through 'dirty' growth.

Fiscal policy experiments

To assess the government's influence on growth and pollution, we simulate two different energy policy schemes. The first one consists of a proportional tax on the consumption of carbon-based fuels. The tax is recycled as a subsidy on non-carbon-based fuel consumption. This generates a substitution effect away from carbon-based fuels and towards non-carbon-based fuels. We introduce the policy in the year 100.[16] Figure 13.8 shows that the policy does increase environmental quality. The introduction of the tax/subsidy leads to an immediate fall in carbon-based fuel consumption, and environmental quality is immediately improved. As the economy grows, carbon-based fuel consumption grows along, but it remains far below its base run value, while environmental quality remains higher than in the base run.

Figure 13.9 shows the policy's impact on real disposable income (RDI). There is an initial drop in output, which is not surprising because the tax/subsidy creates a distortion in the final output sector. However, this output loss is eventually overcome: In the year 166, output under policy 1 surpasses the base run value. From that year on, output is actually higher than in the base run. Thus, the policy may lead to a long run double dividend with higher RDI and lower pollution. To determine the effect on welfare, however, one would have to weigh the present value of the short run output loss against the present value of the long run output gain, and the outcome will then depend on a subjective discount rate.

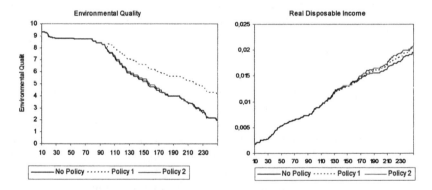

Figure 13.8 Policies and environmental Figure 13.9 Policy impact on real
            quality   disposable income

Figure 13.10 shows why there is an increase in output in the long run. The number of GPT's grows faster under policy 1, so that at the end of the simulation period there are now 16 GPT's instead of 14. The reason for the faster arrival of GPT's is that although the policy targets only the fuel market, it also has an effect on the allocation of R&D resources. By lowering the relative price of non-carbon energy, the policy raises the value of a patent on a non-carbon technology. The present value of these patents rises relative to that of the carbon-technology patents. Since the dominant C04 offering the most applied research opportunities is carbon-based, there will be a larger incentive to do basic R&D on finding the next non-carbon based GPT. Thus, there is a shift from applied R&D to basic R&D, and we have already shown that due to the externality of basic research, there is generally too little basic research in equilibrium. Therefore, the policy-induced shift towards basic R&D leads to a higher long run growth rate.

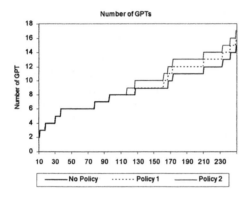

Figure 13.10 Policies, GPTs and output

Realising that the increase in growth was only possible because the market failure in the R&D sector was partially overcome, one might be tempted to tackle the problem at its root by subsidising basic R&D directly. Therefore, our second policy experiment is the same carbon tax as before, but now we use the tax revenue to finance a subsidy on basic R&D. Figure 13.10 shows that this policy is indeed very successful in terms of growth. There is still the initial output loss, but it is overcome in the year 118, much sooner than under policy 1. Thus, in a net present value analysis, this policy has a much better chance of yielding a positive result than policy 1 did. The reason for this increase in growth is that the reallocation of R&D activity is much larger, and so the arrival rate of new GPT's is sped up considerably.

Although policy 1 is highly beneficial for growth, it is less effective in reducing carbon emissions, for two reasons. First, the change in the relative price of non-carbon-based fuels is smaller. Under policy 1, the price effect of the tax on carbon was reinforced by the subsidy on non-carbon. Now there is only the tax effect, because the subsidy on non-carbon-based fuel is replaced by another subsidy. With a smaller relative price change, the substitution effect is weaker. The second reason is that the increase in growth again raises the demand for both carbon and non-carbon-based fuels. As a result, from the year 174 on, carbon-based fuel consumption is actually higher than in the base run, and environmental quality is lower than in the base run.

The potential welfare effects of the different policy schemes depend crucially on the assumptions we make about preferences, especially with respect to the elasticity of substitution between consumption and environmental quality. However, in the long run, both policy schemes are bound to be welfare-improving. In the case of policy 1, this positive effect comes mainly from the cleaner environment, but also from the increase in final output. In the case of policy 2, the positive welfare effect comes from higher output, which offsets the increase in pollution, the more so if environmental quality and consumption are good substitutes. Second, with both policy schemes there may be an initial negative impact on welfare if substitutability is high. This is intuitively plausible, because the initial output loss itself that is caused by the introdution of the tax is welfare-decreasing. But over time, both policy schemes increase growth, thus raising welfare again. In order to get an impression of the robustness of the results, we have run a number of different simulations. It turns out that policy 2, the subsidy on basic R&D, is always growth increasing, because it provides a remedy for the market failure in the R&D sector. Of course, the effect of the policy also depends on the size of the tax/subsidy. If we set the subsidy on basic R&D extremely high, we will at some time reach the point where basic R&D is actually too high, and growth is slowed down again. Hence, there must be an optimal subsidy that exactly equates the social marginal benefits of basic and applied R&D. The exact value of this optimal subsidy is difficult to determine in a model of this complexity, however. The ultimate reason for this is that the arrival of new GPT's and the subsequent expansion of these GPT's is governed by random events, which would allow us to equate these marginal social benefits only in an average sense.

In addition to this, the model generates path dependencies which, in combination with the randomness referred to above, make it impossible to define generic policy prescriptions even in an average sense. To illustrate the point, in the scenario described above, a carbon-based GPT happened to be the dominant one for most of the simulation period. Under these circumstances, policy 1 will induce researchers to move away from applied R&D into basic

R&D, because they hope to discover a new GPT that is based on non-carbon-based fuels, and we have shown that reallocating resources towards basic R&D speeds up growth. If, on the other hand, a non-carbon GPT happens to be dominant, the results are exactly reversed: resources shift from basic R&D towards applied R&D on the non-carbon GPT, and consequently growth slows down. Therefore, the growth effect of policy 1 depends crucially on the existing technology structure.

To check our intuition behind the effects of policy 2, in which tax revenues are used to subsidise basic R&D, we have run simulations with yet another policy scheme in which applied R&D is subsidized. This policy 3 is never the optimal one, because it will yield the same substitution effect as policy 2 (and a smaller one than policy 1), but it exacerbates the misallocation between basic and applied R&D, and its effect on growth will generally be negative. The ultimate choice between policy 1 and 2 depends on the preferences that are assumed. In general, however, we can state that if consumption and environmental quality are regarded as good substitutes, policy 2 is likely to be optimal because it speeds up growth, albeit at the cost of higher pollution. But if they are considered to be poor substitutes, policy 1 is likely to be preferable, because it is more effective at bringing down pollution.

Finally, it should be mentioned that the effects of policies in this model are usually not very long-lasting. With this statement we mean that if a policy is removed – even after a very long time period – the economy tends to jump back towards a development path very close to the one it would have followed without the policy. This phenomenon is due to the 'putty' capital we have assumed, and that is a standard feature of many endogenous growth models. If a policy measure changes the relative price of any fuel, the capital stock is completely free to adjust instantaneously. In reality, capital is often 'clay *ex-post*'. That is, if the price of one type of fuel rises, it is not possible to immediately retrofit all engines to use the other type of fuel. In that case, the effects of policies tend to be more long-lasting than in the current model, thus emphasising the importance of the actual timing of policy measures in the face of limited substitution possibilities *ex post*.

CONCLUSIONS

We have constructed a model in which different GPT's can exist next to each other. These GPT's are based on a highly successful core technology that is extended with a number of peripheral technologies. The model creates long run growth with a Schumpeterian flavour, because the introduction of a new core technology (a 'radical innovation' in Schumpeterian terms) is followed by a quick succession of peripheral technologies ('incremental innovations') that give rise to alternating phases of fast and slow growth, thus capturing many of the stylized facts that we observe in actual economies over the long run. Each GPT is based on either carbon fuels or non-carbon fuels. In such a framework, policies meant to discourage the use of carbon fuels can have a large impact on the allocation of R&D resources.

Using this model, we have examined the impact of different policy scenarios that all include a tax on the consumption of carbon-based fuels, but have different tax recycling schemes. The results for each policy scenario differ considerably, indicating that it is very important to know how a carbon tax is going to be recycled. Moreover, the effect of such a policy also depends crucially on the existing technology structure. If the dominant GPT is carbon-based, a carbon tax will speed up the search for new GPT's, thereby stimulating growth. But if a non-carbon-based GPT dominates, the carbon tax might result in excessive applied R&D on that GPT, thus delaying the arrival of new GPT's and reducing growth.

We have shown that if the competitive allocation of R&D resources is inefficient, policy may increase the growth rate. In that case, the carbon tax alone is a second-best solution and should be combined with a policy that addresses the inefficiency in R&D allocation. This inefficiency results from a non-optimal allocation of resources between basic and applied R&D. Therefore, if growth were the primary policy objective, it would be easiest to tax applied R&D and subsidize basic R&D, without creating any distortions in the energy market. If one wishes to promote a carbon tax on the grounds of R&D efficiency, one would have to show that R&D on carbon technologies is inefficiently high. In fact, our model provides potential support to such an argument.

The scenario we have chosen for the simulation analyses is one where a carbon-based GPT happens to be dominant. Therefore, the economy is highly polluted, and environmental quality is the 'scarce' good. If, however, the economy was in a situation where either output is low or 'clean' technologies dominate, the policies may have fundamentally different effects. Therefore, there may not be a universally optimal policy. It might be optimal, for example, to employ policy 2 at low income levels to speed up growth, and then, at higher income levels, to move towards policy 1 in order to curb carbon-based fuel consumption.[17] More simulation experiments may yield further insights into these matters.

NOTES

1. See, for example, the February 2005 issue of *Scientific American*, or the August 2004 issue of *Physics Today*, both of which describe the detrimental effects of (accelerated) global warming on the Arctic regions.

2. The current model is an extension of the GPT-model described in van Zon *et al.* (2003), hereafter called the ZFK model, with energy as a factor of production. Here we describe the general features and working principles of the extended model. For more details we refer to the description of the ZFK-model.

3. Cf. Ethier (1982).

4. For example, we assume that the real interest rate is constant. However, we are now working on a simplified general equilibrium version of the extended ZFK-model from which these simplifying assumptions are removed.

5. We assume that the most important GPT extensions are made first, which leads to decreasing returns to R&D.

6. The decreasing returns to variety are a relative novelty that hinges on the existence of asymmetries between individual intermediates. Such asymmetries are also present in van Zon and Yetkiner (2003).

7. There are several reasons why an innovation may not grow into a full-fledged GPT. This may be due, for example, to high costs or a limited scope of expansion, or low intrinsic productivity of the 'core-idea'. For more details, see van Zon *et al.* (2003).

8. For further (non-R&D related) details see van Zon *et al.* (2003).

9. The latter characteristics are the inherent productivity of the next GPT, the user costs of the peripherals, the scope for extension of the core, associated applied research productivity and research opportunities. It should be noted here that the scope for extension directly interacts with the Love of Variety feature of the multi-level Ethier production we have used, as a low scope for extension diminishes the impact of Love of variety on output growth. For more details on this interaction, as well as the exact parameterisations of the characteristics involved, see van Zon *et al.* (2003).

10. In our model, however, the riskiness of an activity does not directly influence decision making *ex ante* (although it could *ex post* through temporary lock-in, for example).

11. It should be noted that the curve for applied R&D is really the horizontal summation over the VMP-curves associated with applied R&D in all currently existing GPT's. In addition, the price term present in the VMP-curve is the expected present value of the profit stream associated with the GPT component under consideration (cf. Aghion and Howitt (1992)). For further details, see van Zon *et al.* (2003).

12. This adds a Leontieff layer at the component level to the existing multi-level Ethier production structure.

13. The names Ci and Nj refer to carbon-based GPT i and non-carbon-based GPT j.

14. See van Zon *et al.* (2003) for the details regarding the calculation of this contribution.

15. This is because the pervasiveness of a GPT depends in large part on its performance relative to other existing GPT's. If, for some historical reason, the latter are not very productive, then even a mediocre GPT may become pervasive, in accordance with the saying 'in the kingdom of the blind, the one-eyed man is king'.
16. We use the first 100 years to get rid of the influence of the initial values of the stock variables in the model.
17. This is reminiscent of the environmental Kuznets curve. See, for example, Smulders and Bretschger (2000).

REFERENCES

Aghion, P. and P. Howitt (1998), *Endogenous Growth Theory*, Cambridge MA: MIT Press.

Aghion, P. and P. Howitt (1992), 'A model of growth through creative destruction', *Econometrica*, **60**, 323-351.

Bresnahan, T. and M. Trajtenberg (1995), 'General purpose technologies – engines of growth', *Journal of Econometrics*, **65**, 83-108.

Ethier, W.J. (1982), 'National and international returns to scale in the modern theory of international trade', *American Economic Review*, **72** (3) (June), 389-405.

Francois, P. and H. Lloyd-Ellis (2003), 'Animal spirits through creative destruction', *American Economic Review*, **93** (3), 530-550.

Freeman, C. and C. Pérez (1988), 'Structural crises of adjustment, business cycles and investment behaviour', in Dosi *et al.* (eds), *Technical Change and Economic Theory*, London: Pinter Publishers.

Helpman, E. (1998), *General Purpose Technologies and Economic Growth*, Cambridge MA: MIT Press.

Matsuyama, K. (1999), 'Growing through cycles', *Econometrica*, **67**, 335-347.

Rifkin, J. (2002), *The Hydrogen Economy*, Cambridge: Polity.

Romer, P. (1990), 'Endogenous technological change', *Journal of Political Economy*, **98**, 71-102.

Smulders, S. and L. Bretschger (2000), 'Explaining Environmental Kuznets Curves: How Pollution Induces Policy and New Technologies', *Discussion Paper 12/00*, Ernst-Moritz-Arndt University of Greifswald.

Zon, A. van, E. Fortune and T. Kronenberg (2003) 'How to sow and reap as you go: a Simple Model of Cyclical Endogenous Growth', *MERIT Research Memorandum*, 2003-026.

Zon, A. van and I.H. Yetkiner (2003), 'An endogenous growth model *à la* Romer with embodied energy-saving Technological Change', *Resource and Energy Economics*, **25** (1), 81-103.

Index